The Logical Foundations of Scientific Theories

T0141307

Science is not just a collection of laws, a catalogue of unrelated facts. It is a creation of human mind, with its freely invented ideas and concepts. Physical theories try to form a picture of reality and to establish its connection with the wide world of sense impressions. Thus the only justification for our mental structures is whether and in what way our theories form such a link.

—A. Einstein & L. Infeld

No, no, you're not thinking; you're just being logical.

—N. Bohr (to Einstein)

This book addresses the logical aspects of the foundations of scientific theories. Even though the relevance of formal methods in the study of scientific theories is now widely recognized and regaining prominence, the issues covered here are still not generally discussed in the philosophy of science. The authors focus mainly on the role played by the underlying formal apparatuses employed in the construction of the models of scientific theories by relating the discussion with the so-called semantic approach to scientific theories. The book describes the role played by this metamathematical framework in three main aspects: considerations of formal languages employed to axiomatize scientific theories, the role of the axiomatic method itself, and the way set-theoretical structures, which play the role of the models of theories, are developed. The authors also discuss the differences and philosophical relevance of the two basic ways of axiomatizing a scientific theory — namely, Patrick Suppes's set-theoretical predicates and the "da Costa and Chuaqui" approach. This book engages with important discussions of the nature of scientific theories and will be a useful resource for researchers and upper-level students working in the philosophy of science.

Décio Krause is Professor of Logic and Philosophy of Science in the Department of Philosophy of the Federal University of Santa Catarina, Brazil. His works deal with the logical foundations of science, mainly involving the logical and metaphysical discussions about identity and individuality of quantum entities and the applications of non-classical logics to sciences.

Jonas R. B. Arenhart is Associate Professor at the Department of Philosophy of the Federal University of Santa Catarina, Brazil. His main research interests are logic and philosophy of logic, philosophy of science, the metaphysics of science, and foundations of science.

Routledge Studies in Philosophy of Mathematics and Physics

Edited by Elaine Landry, University of California, Davis, USA and
Dean Rickles, University of Sydney, Australia

The Logical Foundations of Scientific Theories

Languages, Structures, and Models

**Décio Krause and
Jonas R. B. Arenhart**

Routledge
Taylor & Francis Group

LONDON AND NEW YORK

First published 2017
by Routledge

2 Park Square, Milton Park, Abingdon, Oxfordshire OX14 4RN
52 Vanderbilt Avenue, New York, NY 10017

Routledge is an imprint of the Taylor & Francis Group, an informa business

First issued in paperback 2019

Library of Congress Cataloguing-in-Publication Data
Names: Krause, Decio. | Becker Arenhart, Jonas R.
Title: The logical foundations of scientific theories : languages, structures, and models / by Dâecio Krause and Jonas R.B. Arenhart.
Description: New York : Routledge, 2017. | Series: Routledge studies in the philosophy of mathematics and physics ; 1 | Includes bibliographical references and index.
Identifiers: LCCN 2016006566 | ISBN 9781138684492 (hardback : alk. paper)
Subjects: LCSH: Science—Theory reduction. | Science—Philosophy. | Logic, Symbolic and mathematical. | Science—Methodology.
Classification: LCC Q175 .K87194 2017 | DDC 501—dc23
LC record available at https://lccn.loc.gov/2016006566

ISBN: 978-1-138-68449-2 (hbk)
ISBN: 978-0-367-88968-5 (pbk)

Typeset in Sabon
by Apex CoVantage, LLC

To my girls Mercedes, Heloísa, and Olívia (DK)

To mom, Dai, and Carol (JRBA)

Contents

Figures

Preface

It is often said that scientific knowledge is, largely, conceptual knowledge. We approach knowledge (in science) with the help of concepts we elaborate on in order to study a certain domain we are interested in. But we would like to say that scientific knowledge involves a little bit more: we approach knowledge through *structured* concepts. Structures are what we use to approach scientific knowledge. It is with the help of structures that we organize the concepts we use in our scientific activity. With structures, the scientist may have a list of domains of objects she is interested in, a list of relations, operations, and distinguished elements that are relevant for her approach; the interrelations between the distinct concepts become explicit. So a structured account of scientific knowledge seems to capture much of what is generally said about scientific knowledge being conceptual.

We remark that the notion of structure we are using here is very general. Even in psychoanalysis, the specialist uses structures of a kind, say, when she organizes her knowledge with concepts such as (for a Lacanian) the unconscious, repetition, transference, and drive. In this book, we shall be concerned with mathematical structures — that is, those involving concepts of mathematized sciences, mainly physical sciences. We hope that our investigation will throw some light on the nature of scientific theories, mainly on the relation between structures and the linguistic apparatuses employed to describe such structures.

In more precise terms, what is a structure in the sense that we use this term here, and how does it enter into the discussion and into the elaboration of a scientific theory? We shall give precise definitions of these ideas and, more importantly, we shall pay attention to the mathematical framework *where* the notion of structure is constructed. It is important to notice that in the literature there are several approaches to structures, ranging from their mathematical definitions (Bourbaki) to their use in mathematics (Bourbaki, again) and in more general scientific contexts (Carnap, Suppes). But no discussion was provided until now on the kind of mathematical apparatuses a structure depends on and how it influences the idea of a scientific theory that depends on such structures. If a structure is, as we usually learn in our first-order logic courses, a tuple of the form

$\mathfrak{A} = \langle D, \{a_i\}, \{R_j\}, \{f_k\}\rangle$, with i, j, and k ranging over certain sets of indices, then we need to use set theory. But which set theory? We rarely discuss questions like this and others related to them: Can the domain D *really* be considered a set (a collection of *distinct* elements)?[1] And what about the individual constants a_i? Do they *really* name some of the individuals of D? Are all of them nameable? What about the relations R_j? Are they just relations relating elements of the domain? But what if the structure is a topological space, a well-ordered set, or something else, as almost all structures that *model* scientific theories are, which are not *first-order structures* like the one exemplified earlier? Even if we say that we are reasoning within a specific set theory, what about the structures that model the set theory itself? Well, one could say we can always appeal to informal set theory, where (almost) *everything can be done*. This is true, mainly because informal set theory is inconsistent.[2] So should we regard our best scientific theories to be grounded on an inconsistent mathematical basis? From the point of view of the logical foundations of science, this sounds undesirable. As philosophers, we should look for suitable frameworks for grounding our best theories. Except, of course, if the theory is to be inconsistent. But in this case, we need to change logic accordingly.

In addition, we may ask why use set theory and not higher-order logics or category theory? Are these alternatives not good enough to provide a better ground to act as pillars of scientific theories? Of course this also needs to be answered. This book will not address these questions in detail. We simply assume the standard framework most philosophers of science usually make use of *when* they say something about the subject — namely, set theory. In fact, the semantic approach to scientific theories, a paradigm we are still living in, considers scientific theories in terms of their *models*, which, pushing the issue a little, are set-theoretical entities. But the semanticists (van Fraassen, Giere, Suppes, and many others) do not speak about the set theory they are working in. Suppes explicitly regards the metamathematical framework as being *informal* (not axiomatized) set theory, as is well known, but if asked to be more precise (as one of us has required him to be), he would say (as he did) "choose one you find suitable for expressing the mathematical concepts you will be in need of" (this informal conversation took place in Florianópolis in 2002). Okay, we could be happy with this answer, but we are not. Ever since the 1990s, one of us, and now both of us, has investigated standard set theories on what regards the notion of an individual. Since there is a reasonable interpretation of quantum objects in terms of *non-individuals* (see [FreKra.06]), we think that the best way to express *quantum semantics* is not in terms of sets, but by using collections that may comprise indiscernible elements called *quasi-sets*. The theory of these entities is not the subject of this book, but has motivated us to look to the 'standard' set theories used in science. Indeed, physics uses concepts such as vectors, matrices, derivatives, differential equations, tensors, topological spaces, and so on — all of them described in terms of sets. Thus, as part of a major project, we have made the present study we now offer to the general reader interested in the philosophical and logical foundations of science, mainly of physical theories.

We owe much to the approach of Bourbaki's notion of species of structures, mainly in the formulation given by Newton da Costa. In due time, we shall make all the relevant references. We had planned to write a book discussing the nature of the metamathematical framework that underlies the most relevant physical theories, but the project became so wide that we needed to shorten it and keep with a more limited goal. This general plan was not put aside. Thus this book may be seen as a first contribution to the analysis of the underlying mathematics of scientific theories. Indeed, it would require a careful analysis of all physical concepts to see which kind of mathematics they seem to demand; for instance, it is usually said in informal conversations that the denumerable version of the Axiom of Choice is enough for physics, but as far as we know, no proof of this fact has been provided, and perhaps it cannot be done since we don't know the physics that will be developed in near future. The study could, of course, be restricted to known physical theories. But this is not what we intend to do here, for we are not concerned with the history of science. Our aim is to offer to the reader an overview of the notion of mathematical structure that underlies the usual physical theories. Thus this book is organized as follows.

The first chapter rehearses most of the classical opposition between the so-called *syntactical* and *semantic* approaches to scientific theories. We briefly characterize the two approaches and their main features, pressing on the alleged virtues of the semantic approach over the deficiencies of the syntactic approach. As it is widely known, and we discuss in the chapter, language is one of the central points of concern: the syntactic approach is said to fail due to its heavy reliance on language, while the semantic approach succeeds because it may characterize theories as language-free. We present how the current debate has shifted from such a concern over language and how both approaches may be seen as somehow complementary. In particular, we advance the thesis present throughout the book that the characterization of theories most appropriate depends on our purposes. That is, how we present a theory depends on what we need. In most cases, both a linguistic approach and the use of models are helpful for distinct purposes and both are needed for metamathematical studies of scientific theories.

With the main objection that the use of a specific language may cause trouble in the characterization of a scientific theory, in chapter 2, we discuss our first tool: the axiomatic method. In fact, a theory may be characterized by the linguistic formulation of its axioms or else by a class of models, models of the axioms. So axioms are the first fundamental tool for the relevant study of foundations. We keep the chapter self-contained by briefly presenting the history of the method and its evolution. In the last case, the evolution comes in two senses: in one direction there is a growing abstraction in the method from the nineteenth century. In another sense, there is, also from the nineteenth century on, the recognition that this is a useful tool for conceptual clarification for empirical sciences such as physics (obviously, the method was a model and an ideal for philosophers ever since its inception in the antiquities).

We continue our investigation on the axiomatic method in chapter 3. We exemplify the use of the axiomatic method to describe the theory that will serve as a mathematical basis for our discussion of structures and also will work as a framework inside which axiomatization of most scientific theories takes place. The theory in question is the famous Zermelo–Fraenkel set theory. Set theory is the mathematical basis in which models as set-theoretical structures are developed, so it is useful, to keep the book self-contained, to have a good look at how it is developed. Also, it is a case of a formal theory, so it continues the discussion given in chapter 2. Given that set theory is a *sui generis* theory, due to the fact that its models cannot be developed inside set theory itself (unless it is a stronger set theory, as we shall discuss), we also present a brief discussion on the nature of models of set theory. Models of set theory play a much greater role in the study of scientific theories than most people think: the nature of the model of set theory being employed influences the nature of the models of scientific theories that are being developed inside that set theory. We discuss such issues in this volume, and even though we cannot develop the whole point in this book, this is a first approach to the subject.

In chapter 4, we present some criticisms advanced against the axiomatic method. We argue for its usefulness in the development of science and try to show that the main criticisms are not effective. In particular, criticisms are presented against the Suppesian approach to axiomatization, an approach we shall present in detail in the next chapter. We argue that the criticism in this case is, in general, misguided, confusing a general presentation of the axiomatic method with a particular use made of it; that is, the critics attack the fact that the method was employed only to deal with classical particle mechanics and that this restriction renders it useless. However, as a presentation of a method, it is perfectly legitimate, and extensions may still be provided for other areas of physics. That is, the method is not restrictive and limited because it is being used to deal with a limited field of physics.

Chapter 5 addresses the issue of axiomatization of empirical theories. We are mainly concerned with two approaches to axiomatization: one by da Costa and Chuaqui and another by Suppes himself. Here our previous discussion on the use of language and on the nature of models begins to enter our analysis. As we argue, da Costa and Chuaqui axiomatize a theory by presenting its postulates in a formal language. The models of the theory thus framed are in fact models in the style of Tarski. We keep the text self-contained and present details about the construction of structures and formal languages. We also present as an example the axiomatization of classical particle mechanics in the style of da Costa and Chuaqui; this requires that every mathematical step theory be axiomatized too. In opposition to that approach, Suppes's axiomatization in set theory assumes not a specific formal language for a theory, but rather set theory itself as the language in which the theory will be axiomatized. This is in straight contrast with the da Costa–Chuaqui approach. We exemplify how this restriction to the language of set theory increases the power of the axiomatization and how it facilitates axiomatization. Examples are also provided. So while da Costa–Chuaqui axiomatization is more apt for metamathematical studies, Suppes's approach focuses on the easiness of application.

Finally, in chapter 6, we develop a discussion on the nature of models for each of the kinds of axiomatization developed in the previous chapters. As we mentioned, da Costa–Chuaqui axiomatization is a typical logical axiomatization, with models playing the role of providing an interpretation for the axioms. In this sense, a class of models is a class of models in the Tarskian sense, with an interpretation of the language of the theory. This provides for nice opportunities to apply the resources of higher-order model theory to scientific theories in particular, the theory of definition presented briefly in chapter 5 may contribute with a possible solution to the problem of equivalence between theories as formulated in chapter 1. We address these issues in this chapter. Suppes's approach, on the other hand, is not a legitimate model approach, as it were. The models for an axiomatic system in the style of Suppes are not taken in the same sense as Tarski's models; the notions of interpretation and truth are absent. We discuss such issues in this chapter and how one can account for some set-theoretical structures being the 'models' of a theory. We also discuss how the models of the set theory being employed are relevant to the nature of the models that characterize a theory. Even though our discussions are not fully developed, they set the stage for further investigations and developments, which we hope to conduce in our investigations to follow.

The reader may take this book as a first step, a reopening of the field, an attempt to bring metamathematical issues back to the study of scientific theories in particular, and philosophy of science in general. We hope this rather initial and inconclusive investigation will inspire others to follow along similar lines.

Florianópolis, December of 2015
The authors

NOTES

1. A pioneering questioning in this sense appeared in [DalTor.93]; see also [FreKra.06].
2. Inconsistent in an intuitive sense.

Acknowledgments

We would like to acknowledge the influence and help of several people. With the risk of leaving someone out, we would like to thank Newton da Costa for his influence, Otávio Bueno for constant discussion, Steven French, and Fernando T. F. Moraes. We must not forget to mention our students Conrado Emerick and Lauro de Mattos Nunes Filho, who had the patience to take our courses on scientific theories while we were still working on the manuscript. Special thanks to Dean Rickles and Elaine Landry, the editors of the series *Philosophy of Mathematics and Physics* for the kind invitation for us to write this book. We also would like to give special thanks to Sophie Rudland and Chris Mathews for the kind correspondence, for the attention, and for the guidelines in preparing this manuscript.

1 The Quandary on the Characterization of Scientific Theories

It is perhaps a common place to say that the debate on the nature of scientific theories in contemporary philosophy of science has mainly revolved around the syntax-semantics debate. During the first half of the twentieth century, when the Logical Empiricist program flourished, a version of the *syntactical approach* ruled. During the second half of the twentieth century, after the development of model theory by Tarski and the decline of the Vienna Circle, the *semantic approach* was developed and still forms the actual paradigm. However, a closer look would reveal that apart from the story of a heated dispute, there was not really a debate: after the decline of the *Received View*—as the approach by the Logical Empiricists to scientific theories was called—almost no one else was really promoting that kind of view, so it is strange that philosophers of science adhering to the semantic approach took as part of their tasks to criticize what they took to be the syntactic approach.

F. Suppe's preface and afterward in [Sup.77] illustrate the situation quite clearly; Suppe's volume is a collection of papers presented at Symposium on the Structure of Scientific Theories held at Illinois in 1969. The call for the symposium urged participants to promote new views on the structure of scientific theories, given that the Received View was considered as largely inadequate (again, see the preface to [Sup.77]). Facing the ruin of the only unitary framework available for philosophical work on scientific theories, philosophers of science were looking for a substitute. As Suppe [Sup.77, p.618] has put it, after such a successful debunking of the Received View,

> the result was widespread confusion and disagreement among philosophers as to what the main problems in philosophy of science were, how they should be approached, and what would constitute acceptable solutions to them. The 1969 Illinois Symposium on the Structure of Scientific Theories [...] occurred in the midst of this disarray, and thus provides a particularly vivid account of a discipline in search of a new direction.

So, it seems, a framework for scientific theories is required in order for the main problems in the philosophy of science to be clearly stated and investigated. The Received View furnished such a framework, and in its absence, no consensus could be achieved about even what were the main problems in the philosophy of science. A replacement was being sought and was found in the new semantic approach. The later did not seem to fall prey to the difficulties that plagued the Received View. However, the dialectics of the debate seem to put much emphasis on that difference. A cursory look in the literature will give the reader the impression that there is still a debate; adherents of the semantic approach still seem to feel the need to constantly review criticisms of the Received View and promote the advantages of the semantic approach on the treatment of those difficulties.

Given that still prevailing agonistic spirit, the central issue concerns whether scientific theories (specifically, empirical theories) are more properly characterized as syntactical entities, in terms of formal languages and sets of axioms and inference rules expressed in such formal language, or as semantic entities, in terms of classes of models and/or structures (there is disagreement over whether structures and models are the same kind of thing and about which is more appropriate for the semantic approach, as we shall see soon). It is generally agreed that the semanticists have won the battle, and the semantic approach is now considered the new orthodoxy [Con.06]. Now, as we have remarked, it should come as no surprise that the semantic view established itself, given that the syntactic view was seen as unable to deal with its criticisms even before the semantic approach clearly emerged. Along with the rise of the new orthodoxy, a less rigorous and less formal-friendly mood has dominated the philosophical studies of scientific theories.

However, apart from how the story is usually told, new studies on the Received View are emerging, and along with them, a more faithful understanding of the characterization of scientific theories by the Logical Empiricists is being achieved. With the publication of such works, we start to understand that the debate is presented in such a way that is not favorable to the syntactical approach; in fact, the Received View is generally presented as a caricature of a highly naive and implausible view (see [Lut.12], [Hal.15], and the references in those works). The syntactic approach, mainly *identified* with the approach advanced by some members of the Vienna Circle such as Carnap and Hempel, was criticized in almost every aspect. As it was characterized by its opponents, it really did suffer from profound difficulties as an approach to scientific theories. Perhaps the heavier criticisms seem to be those accusing it of too radical deviance of actual scientific practice, mainly due to its heavy reliance on first-order logic and axiomatization. As an account of scientific theories, the Received View failed badly by distancing itself from real science and by relying so heavily on formal tools and techniques (or, at least, so the argument goes).

The semantic view, on the other hand, seemed to be completely different from the syntactical view in those aspects, keeping close to actual scientific practice and not requiring that scientific theories be formulated in any specific language. In particular, this last feature was erected as a great virtue of the view and was defended by van Fraassen and his followers (see for instance [vanF.89, pp.221, pp.225–6]). The so-called model revolution initiated by Patrick Suppes in the '60s would be reduced to nothing if language were allowed to play a substantial role in the formulation of a theory (the claim is not from Suppes himself, but see [Sups.60, Sups.67] and [Mul.11, sec.6]). However, as we shall discuss in what follows, it is perhaps this sole requirement of being 'language-free' that makes the semantic approach almost senseless, while at the same time it is this requirement that allegedly marks a radical divide between both approaches nowadays. Leave that requirement out and we have a position that can, perhaps, be made compatible with a syntactical approach too.

We shall not attempt to present here a revision of the literature about the whole dispute. However, given that the following chapters will deal with issues that are related to both the semantic and the syntactic approaches, and references shall be made to those approaches, we shall give here to the reader a brief summary of the debate and present reasons for it seemingly going out of the tracks. Our aim is not to promote one of the approaches as superior, but rather to argue that scientific theories may be profitably studied by the philosopher and by those interested in foundations from many distinct points of view. Instead of a competition between distinct approaches, we propose that they do complement each other. This kind of claim will involve another revision of the attitude toward the relation between theories in real scientific practice and our rational reconstruction of them for philosophical purposes. As we shall argue, our constructs may employ distinct technical resources, and it is not clear that they should reproduce in every detail their informal counterparts. Perhaps formalized theories (be it in a formal language or some set theory) gain a life of their own, helping us to understand their informal counterparts; that is their purpose.

So, for the moment, we shall briefly review the main features of the Received View (and mention syntactic approaches) and of the semantic approach. This, we hope, will show how features of both approaches are employed in the next chapters with the aim of logical analysis of scientific theories. It should be clear that an emphasis on the *identification* of theories with something seems responsible for much of the problems.

1.1   THE RECEIVED VIEW AND SYNTACTICAL APPROACHES

The traditional presentation and criticisms addressed to *the syntactic view of theories* should more properly be confined only to what we have called the *Received View* on scientific theories. Recall that the Received View is the view on scientific theories developed by members of the Vienna Circle, mainly Carnap and Hempel (the classical exposition may be found in the introduction by F. Suppe to the volume [Sup.77]). While the Received View is clearly a syntactical view, it is not the only incarnation a syntactical view may have. The Received View is clearly a syntactical view in its advocacy of formalizing theories by employing a formal language and presenting it as an axiomatic calculus, but the fact that a theory is characterized as a formal system does not by itself imply that the characterization follows the Received View; as we shall argue briefly, there is much more to the Received View than to a syntactical approach in general.

Traditionally, the Received View is characterized by requiring that theories be framed according to the following features:

Language: There is a formal language whose non-logical vocabulary is divided in two parts: V_T and V_O. V_T is the set of theoretical terms, and V_O is the set of observational terms. Notice that there is no restriction on the order of the language or kind of language to be used (it does not prohibit higher-order languages, modal languages, and so on; see for instance [Car.58]).

Logic: The language has a set of logical axioms, giving rise to its underlying logic (in general, type theory; see [Lut.12, sec.2]).

Theoretical postulates: A set of sentences written exclusively in the theoretical vocabulary is selected as the set of *theoretical postulates*, which we shall denote by T.

Semantics for V_O: Observational terms receive an informal semantics relating them to observable objects and events. This is related to the theory of meaning as verification adopted by the Logical Empiricists.

Correspondence rules: A set of sentences relating theoretical terms to observational terms. These provide for a partial semantics for theoretical terms, relating them with immediate experience. We shall call C the set of correspondence rules.

So, characterized by the aforementioned requirements, a theory is individuated by its theoretical postulates and its set of correspondence rules. Let us call, accordingly, TC the resulting theory. It is of particular relevance that by framing a theory with the specific choice of a language, one already commits the theory with being characterized not only by its postulates and correspondence rules but also with a specific vocabulary and language.

Now, each of those features the Received View demanded of a theory received its share of criticism (see again the introduction to [Sup.77] for the classical exposition and criticisms of the Received View; Muller [Mul.11] also presents some of the same points). It is now almost a widespread consensus that some of the demands of the Received View could not be achieved. In particular, there are many aspects that were very closely tied to philosophical theses of the Vienna Circle and which ended up leading the general program of the Logical Empiricists to its collapse in its original form; those aspects inevitably led to problems in the Received View too.

Perhaps the most dramatic example of the mix between philosophical tenets of the Logical Empiricists and their characterization of a theory comes from the peculiar kind of relation between a theory and experience. To mention some of the most prominent ones, there should be a match between observational vocabulary and observable entities. The same should also happen to theoretical vocabulary and non-observable entities. However, it is really difficult to confine the observational vocabulary to refer only to observable entities; depending on the use of quantifiers along with the observational vocabulary, it is easy to produce sentences in the observational vocabulary that do not relate directly with any experience. Also, depending on how one understands the relation of verification by experience, it would be simple to produce examples of experiencing theoretical entities: for instance, by putting our fingers in the electric plug (don't do that), we have an experience of electric current, which should be a non-observable entity.

So the division in the vocabulary and its corresponding division between kinds of entities is very problematic, to say the least (Kuhn, Feyerabend and Hanson, for instance, made much of those difficulties, but we shall not enter such a debate here; see the introduction to [Sup.77, sec.5]). Another very difficult question concerns the precise role of the correspondence rules. Carnap, for instance, had many difficulties in explaining how those rules attribute meaning to theoretical terms. At first, it was required that the correspondence rules had the form of an explicit definition, such as

$$\forall x(Tx \leftrightarrow \Phi x),$$

where T is a theoretical term and Φ is a sentence in the observational vocabulary. That approach had many problems; in particular, it could not be employed to dispositional terms, such as 'soluble' and 'fragile'. For instance, in the case of fragility, it would lead to the implausible

x is fragile $\leftrightarrow$ (if x is severely beaten, then x breaks).

The problem with that approach is that anything that is never beaten ends up being fragile (due to traditional truth conditions associated with the material conditional). So the requirement of explicit definition of theoretical terms had to be relaxed. Less stringent demands were soon provided for, but had their own difficulties. One of the options consisted of correspondence rules providing for conditionals, only partially defining the meaning of theoretical terms. For instance, consider 'fragile' again. Then the proposal by Carnap was the following:

If x is severely beaten, then (x breaks $\leftrightarrow$ x is fragile).

It is difficult to specify what it means to provide only partial meaning to theoretical terms, and related problems with analyticity helped to frustrate the program.

For a related problem associated with correspondence rules, it was argued that demanding that the methods of application of a theory be part of the formulation of the theory leads to inconvenient situations. For instance, whenever a new method of application of a theory is discovered (new apparatuses due to new technology, for instance), the formulation of that theory must be changed accordingly, and we would literally have a distinct theory. Also, requiring that correspondence rules be included in the formulation of the theory seems to put the causal mechanisms of theory test inside the theory itself, which is counterintuitive, to say the least.

In general, the main problem with the Received View concerns its simplified account of the relation between theory and experience. It relied heavily on the empiricists dogma of reduction (sentences must be verified or corroborated one by one) and were meant to introduce in the account of theories the problematic verifiability criterion of meaning. However, notice that one could in fact dispense with those peculiarities of the Received View in alternative forms of the syntactic approach: one needs not adhere to the tenets of the Logical Empiricists to adopt a syntactical view of theories, with a change in the mechanisms of the relation between theory and experience being a nice starting point.

But for our purposes here, much more important than those well-known difficulties in the relation between theory and experience, which most people consider unsurmountable, is the kind of attack leveled against the use of formal languages made by the Received View. This kind of attack is also considered unsurmountable, but the reasons given are not that convincing. Typically, authors complaining about syntactical approaches (and about the Received View, which as we mentioned is generally identified with syntactical approaches) claim that the Received View adopts the thesis that the correct language for formalizing empirical theories is a first-order language, with an accompanying first-order logic (again, see [Sup.77] and also [Mul.11] for that claim). That kind of language is clearly inadequate for most scientific theories, and even for those theories that can be written by using this kind of formalism, the resulting axiomatization is awkward and impractical. In particular, the required mathematics of a scientific theory would have to be wholly axiomatized from scratch, leaving us with so much preliminary work to be done before getting to the theory itself; that requirement not only distances the Received View from actual scientific practice but also would render the whole enterprise useless for any scientific purpose.

As an instance, consider the theory of complete ordered fields, which is part of a good amount of scientific theories employing mathematics. First-order languages are not adequate to axiomatize the theory; in particular, one cannot provide an axiom for the *least upper bound principle*—which characterizes the field of real numbers—as a single formula. In first-order languages, we employ an axiom scheme:

$$(\exists x \alpha \wedge \exists y \forall x (\alpha \rightarrow x \leq y)) \rightarrow \exists z (\forall x (\alpha \rightarrow x \leq z) \wedge$$
$$\forall y (\forall x (\alpha \rightarrow x \leq y) \rightarrow z \leq y)).$$

Here α is any formula of the first-order language for ordered fields, and x, y, and z are variables. The need to use a scheme of axioms and the existence of unintended countable models makes the theory uncategorical. Second-order versions of the principle employ predicate variables in the place of α, and then the axiom can be expressed as a single formula. As it is well known, the second-order theory of complete fields is categorical. The same kind of reasoning may be employed to argue that the theory of well-orderings or cyclic groups are not elementary theories. One cannot obtain the class of well-orderings with a set of first-order sentences; well-ordering is not a first-order class. That is obviously a problem for those limiting themselves to theories with first-order languages (see also [Kun.09, p.89]).

Against that line of criticism, it was argued by Lutz [Lut.12], with appeal to a large textual evidence, that there is no reason to think that Carnap and other adherents of the Received View ever demanded formalization in first-order logics, with its devastatingly impractical consequences. In fact, most of Carnap's work explicitly employs the theory of types, which from a logical point of view is enough to ground not only the required mathematics but also the higher-order concepts involved in science (see for instance [Car.58]). Furthermore, formalization in general, as seen by these authors, did not demand an exhaustive axiomatization of all the mathematics required: it is enough to know that the relevant mathematics can be developed inside type theory. Once it is known how to develop the required mathematics, one may proceed to deal exclusively with the empirical science one is concerned with.

So the objection that the use of first-order languages leads to the failure of the Received View is based on a widespread misunderstanding. In fact, due to the large amount of evidence that the proponents of the Received View (mainly Carnap) did not require exclusive use of first-order languages, it is strange, to say the least, that opponents of the view have focused for so long a time on that aspect.

The objections against languages did not stop at the supposed inadequacies of first-order languages. As argued generally, and made famous by van

Fraassen and Suppe, one of the main problems with the Received View and syntactic approaches in general concerns the use of a specific vocabulary and of axiomatization in such vocabulary. Given that a theory is *identified with its linguistic formulation* (in axiomatic terms), it would result in impossible formulations of the same theory in alternative vocabularies. Now, that claim involves two kinds of presuppositions. One of them is that the Received View did not have means to account for the use of distinct vocabularies in the formulation of a theory. The second concerns the very idea, widespread among critics of the Received View, that a theory *is identified* with its formulation. Do those criticisms really hit the Received View? What are the consequences for a syntactical approach that does not necessarily adopt all the features of a theory as required by the Received View?

As to the first problem, *viz.*, the use of a specific formulation, it seems clear that the Received View need not be that specific. As Halvorson [Hal.15, p.5] and Lutz [Lut.15, p.7] argued, the Received View, and in general any syntactical view, need not be committed to the individuation of a theory as formulated in a particular language: one may clearly accept equivalent formulations of a theory. That requires certainly a notion of *theory equivalence*, but it is clear that such equivalence may be formulated in distinct terms, such as definitional equivalence or equivalence by translations (see [Hal.12, p.191] for definitional equivalence and also [Mul.11, Sec. 6] for translations between distinct formulations). In particular, there is no evidence that the Received View was committed with the individuation of a theory by its vocabulary.

So syntactical approaches in general are not tied to the use of a specific language forever. This should be clear from Carnap's writings, for instance, but was constantly repeated and pointed out as a definitive failure of the syntactical approaches in general. Closely related to this claim, there is another very widespread view about the aims of the Received View (and of the semantic view and syntactic approaches in general): it provides for an explication of the concept of 'theory'. Suppe [Sup.77, pp.57–61], in his famous introduction, identified the whole project of providing a characterization of theories by the Received View as an attempt to provide an explication of the concept of theory in general. Halvorson [Hal.12, Hal.15] also adopts that view, even though his sympathies lie with syntactic approaches.

The main problem with that approach concerns how an explication is to be understood according to Carnap, for instance. An *explication* of a certain concept C, the explicandum, consists of the development of a new concept C^*, the explicatum. The relation between explicandum and explicatum in the Received View, according to Suppe, is such that the explicatum is a *precisification* of the explicandum. In other words, i) every positive instance of a C should also count as a positive instance of C^*, ii) every negative instance of C should also count as a negative instance of C^*, iii) C^* must be more precise than C whenever possible, and iv) C^*

should be fruitful. We can then use C^* to decide about cases in which we ain't sure whether something is a C or not. That view on explication, then, applied to theories, means that the Received View should be seen as advancing an explicatum for the intuitive informal notion of scientific theory, the explicandum.

Now, by understanding explication as a precisification, it is clear that the Received View should fare rather badly. Indeed, almost any approach making use of formal methods will have its problems with such an approach. Obviously, not every scientific theory in the intuitive sense can be axiomatized as a formal system. Also, not every formal system in accordance with the rules of the Received View counts as a theory (it is easy to produce simple counterexamples). So Suppe concludes that the Received View failed in this aspect too, while the semantic approach (which we shall present soon) does much better.

However, it is clear that Carnap and other adherents of the Received View did not adopt a precisificationist view of explication. As argued in [Lut.12, sec.5], Carnap did allow some deviation from common use in explications. The proper rigorous explication of a concept in scientific terms in general advances deviations from common use; in fact, as Carnap himself advances an example, the scientific explication of the term 'Fish' replaces it by another concept, let us call it 'Piscis'. Those concepts do not coincide: while Fish counts whales and seals as its instances, Piscis is much narrower and excludes those cases (See [Lut.12, p.101] for the full quotation and discussion; it is curious that Halvorson [Hal.15] mentions Suppe's account [Sup.77] of explication as illuminating, while at the same time making reference to Lutz's [Lut.12] paper, which has a much more convincing analysis.)

As suggested by Lutz [Lut.12, sec.5], perhaps it would be useful to see the Received View as *a rational reconstruction of particular theories* and not as a general reconstruction of the concept of a theory. As Carnap was aware of, not every theory is so developed that it can be axiomatized and presented in the canonical form proposed by the Received View. So that kind of presentation should be seen as an ideal, a model in which theories developed enough can be presented in order for us to study the empirical meaning of them (or whatever else we are interested in about the theory). So some of the most telling problems for the Received View, say those concerning its adequacy as an explication of the term *theory*, may be overcome by recognizing that such an explication of the general concept of theory is not what is really at stake; rather, what is presented is a method or formal tool for the study of specific theories. It may be useful for certain purposes in certain circumstances, but it may not work for other purposes (see also [DutRec.15] for further relations between explications and the use of formal methods).

Let us now put together some of what the previous discussion has enlightened: a syntactic approach to theories does not need to employ every aspect of the Received View. In particular, those aspects that were already found as lacking, such as the specific relation between theories and experience, the specific division of the vocabulary into theoretical and observational, and the use of rules of correspondence, for instance, need not to be present. Also, the use of some particular language need not commit us to the identification of a theory with its particular formulation: i) on the one hand, the use of formal languages can be made compatible with distinct formulations being required only a notion of equivalence between theories and ii) there is no need of an identification of a theory with anything; in fact, the use of some formal method is not meant to provide for a definition or explication of the concept of a theory. The use of some syntactical method is a resource that philosophers of science and those interested in the foundations of science may deploy for the purposes of their work.

In general, then, we agree with Lutz [Lut.15, p.7] when he characterizes an approach to theories as syntactical if it comprises the following, in general lines:

Formal language: the theory is formulated in some formal language (possibly a higher-order language).

Theoretical equivalence: a relation of equivalence between formulations is provided (so that distinct formulations count as formulation of the same theory).

Interpretation: some sentences are interpreted (to grant that we are dealing with empirical theories).

Notice that the most troublesome aspects of the Received View are absent or need not to be included. In particular, the requirement of interpretation need not be framed in terms of correspondence rules or any other restriction due to the Received View.

1.2   THE SEMANTIC APPROACH

Traditionally, the semantic approach is said to have been born along with the development of model theory by Tarski, in the 1950s (see also [CosFre.03, chap.2–3]). Particularly influential in developing the approach are the works of Suppes from the '50s onward (see for instance [Sups.60, Sups.67]), which we shall follow closely here. Although Suppes may be seen as leading the semantic approach, there is not really a unified view that could be called *the semantic approach* in opposition to the Received View; in particular, depending on how 'models' are understood, the view takes on distinct versions, which vary also as to how models are related to reality and what role language plays in characterizing a scientific theory.

Now, given the existence of disagreement on what precisely models are and how they relate to reality, we shall here adhere to what may be called the hegemonic version of the semantic approach. While this may be controversial, we shall restrict the scope of what we call the semantic view here and focus on the set-theoretical development of models, leaving aside other kinds of models that also feature in those discussions. This should not strike the reader as a big restriction, given that some of the most recent discussions on the syntax-semantics debate were drawn from this kind of approach to model theory.

Following this particular kind of approach to models, most authors consider that the semantic approach involves the following main features:

Models: A theory is seen as a class of models.
Set theoretical structures: Models are set theoretical structures.
Language independence: A theory is independent of language.

As we mentioned before, 'models' here are taken as set-theoretical entities; that is, entities built inside some set theory (typically Zermelo–Fraenkel set theory, as we shall see in the chapters that follow). Following model theory and the work of Tarski, one could think that a model, properly speaking, is an ordered pair

$$\langle D, \rho \rangle,$$

where D is a non-empty set and ρ is an interpretation function relating the non-logical vocabulary of a formal language with the corresponding set-theoretical entities: individual constant symbols are mapped to elements of D, n-ary function symbols are mapped to n-ary functions over D, and n-ary predicate symbols are mapped to n-ary relations over D.

That is a typical kind of description of a model found in the context of the semantic approach. However, notice that some simple shortcomings seem to be forgotten by this specific use of 'model' here. First of all, characterized like that, models are what are usually called 'first-order models': models for languages of first-order, comprising only what we call *order-1* (first-order) structures.[1] By sticking with these models one cannot solve the problem of keeping restricted to first-order languages. Really, there are relevant theories, both in mathematics and in the empirical sciences that cannot be modeled by order-1 structures, requiring structures of higher order. For instance, well-orderings are not classes of models according to that characterization that is, a well-ordering is not a order-1 structure, as is easy to note.

Order-1 structures can be models only for first-order languages and theories written in those languages, which makes the difficulties of first-order languages reflect in the models that may be characterized by those languages. As a second problem, related to the first one, it only makes sense to call a structure a model provided we have a set of axioms that are modeled by the structures in the mentioned set. In fact, models are models of something, and in the present case study, they are models of some axiomatics. In that sense, the semantic approach would predate over a form of syntactic approach: only once we have the axioms of the theory (that is, characterize it syntactically) can we characterize the theory semantically. But, as we shall see, it is not necessary to follow all the steps suggested by the Received View.

Now, even though those problems and limitations are generally overlooked in the literature on the semantic approach, the greatest challenge to the previous characterization of models would be its violation of the *language independence* requirement; that requirement seems to prohibit any characterization of models that involve *interpreting a language*. As van Fraassen [vanF.89, p.366] has emphasized, the advantages of the semantic approach would be lost if language were allowed to play that role. Here it should be understood that those advantages include being free from any specific language; being a structure related with a specific language would render the theory formulation-dependent (or, at least, so the argument goes). In this sense, if that advantage of the semantic approach is to be preserved, it seems that the interpretation function must be left out of the account of a model adequate for the semantic approach. Instead of models in the Tarskian sense, perhaps the semanticist really means another kind of set-theoretical construct when she speaks about models (see also [Mul.11, sec.6] for this kind of discussion).

The most common alternative for the characterization of models in this case is to allow that the relevant 'models' are in fact set-theoretical structures: entities of the kind

$$\langle D, R_i \rangle_{i \in I},$$

where, again, D is a non-empty set and R_i is a family of relations on the elements of D (perhaps comprising also operations and distinguished elements, all of which may be seen as special cases of relations). Here it is important to note that the R_i need not relate only members of D, giving rise to higher-order structures or, as we prefer to call them, order-n structures, with $n > 1$ (we shall discuss such structures in chapter 5). For instance, a group is a set-theoretical structure that may be written as

$$\langle G, \circ, e, - \rangle,$$

where G is a non-empty set, $\circ$ is the composition function, e is the identity element, and—is the inverse operation. Here we have only operations over elements of D, so it is an order-1 structure. A topological space, on the other hand, is a structure $\langle D, \tau \rangle$, where $\tau \in \mathfrak{P}(\mathfrak{P}(D))$. That is, τ is a family of subsets of D, so we have a structure or order greater than one.

Now, if that suggestion is the case (and it really seems to be), then it is difficult to see what the structures thus gathered are models of. One option, as we shall see later (see chapters 5 and 6 on Suppes's predicates) is to say that they are models of a *Suppes predicate*, a set-theoretical formula axiomatizing a theory inside set theory. But then the notion of 'model' is rather different from Tarski's sense of model, and there is no proper model theory, given that they are higher-order structures. In the end, it seems, there is a dilemma to be faced: either the relevant models for the semantic approach are models in the Tarskian sense or else they are not models in that desired sense they are not 'semantic' in the sense of making anything true. In the first case, they are inconvenient because they involve language; in the second sense, it is difficult to say what is really *semantics* in the semantic approach. Anyway, one must put constraints on how the structures are gathered. For instance, in the case of topological space, there is not any structure of the form $\langle D, \tau \rangle$ where τ must satisfy some standard requirements! Language seems important here, as we have mentioned.

Later on we shall argue that the mentioned dilemma is not really a problem if we abandon, as we think we should, the idea that theories are formulation-independent and/or language-free. This requirement, as we shall put it, comes from a desire to *identify* theories with something; our purpose, as we shall argue, is to provide distinct representations of theories, and in that case, it is not a problem that a representation is framed in a specific language (as most representations are).

Before proceeding, let us already fix then the first divide between two notions of 'model' that will be relevant for our further investigations. Both are set-theoretical entities, but they operate in distinct ways. It seems that those adhering to the 'language-free' ideal should submit that the 'models' the semantic approach speaks about are merely set-theoretical structures; there is no language in which those structures are interpreted, that is, sets and relations over those sets (in a general characterization). The best we may have is a 'satisfaction' of a Suppes predicate, an approach to axiomatization that proceeds in the language of set theory itself without requiring a separate formal language for a structure. As we shall discuss later, however, 'satisfaction' here has a different sense than the traditional model theoretic one, as we shall see. Those not adhering to the 'language-free' ideal may still think that the models work just as models in the sense of model theory developed by Tarski (although they may also be models of higher-order languages; see also [KraAreMor.11]). As we shall argue in the next chapters, adherents to the semantic approach were not always clear about this. This distinction was even the topic of a heated debate in the philosophy of science (see [Hal.12] for the first sparkle, and [Lut.15] for the general overview; see also our [KraAreMor.11], where the distinction was already clearly drawn).

These disputes about the proper meaning of models notwithstanding, what most adherents of the semantic approach agree with is that identifying theories with classes of models allows us to avoid most of the problems faced by the Received View. Recall that it is this comparison between the two approaches that sets, in general, the particular dialectics of the debate, putting the Received View (and syntactic approaches in general) in rather unfavorable lights. Due to the alleged language independence of models, theories (taken as a class of models) are independent of their formulations. The same theory can be presented as associated with distinct languages, and none of them are essential to it. Also, there is no need to be worried with formal details of axiomatization in first-order languages; it is enough to specify directly the class of models inside some set theory. In this case, the expressive capability of the semantic approach is said to go much beyond the Received View. Finally, to present a theory as a class of models seems to bring us closer to scientific practice, given that scientists are model builders, not theorem provers (see [Hal.15] for a summary of the typical advantages claimed in the name of the semantic approach).

As we have already mentioned, the semantic approach has been developed in a much more informal way than the Received View (even though it is most of the time said to be a descendent of Tarski). The idea that one can identify a theory with a class of models and that one can even collect a class of models without essential use of a language went largely unanalyzed through many years, until recently Muller [Mul.11] and Halvorson [Hal.12] have taken it seriously and found the idea lacking. The main difficulty, as we shall see, derives from a tendency to identify a theory with a class of models, while providing for no relation of equivalence between classes of models other than identity of classes.

First of all, informal theories have distinct informal presentations, which may seem to lead to distinct classes of models. As an example, Hamiltonian mechanics and Lagrangean mechanics seem to be the same theory, but are certainly not represented by the same class of models; they even employ distinct mathematical apparatuses. Something similar holds to Schrödinger's and Heisenberg's versions of quantum mechanics. In the absence of a criterion of theoretical equivalence, there is nothing in the semantic approach that allows us to grant that intuitive equivalence. In fact, some philosophers adhering to the semantic view have employed such formal dissimilarities to argue that Hamiltonian and Lagrangean versions of classical mechanics are not the same theories (see [Hal.12, p.187] for the discussions and references). So from an intuitive point of view, the semantic approach seems to fail in individuating theories. This, as we mentioned, seems to be a result from bare identification of an intuitive theory with its class of models.

For another main difficulty, it is clear that even some simple theories such as group theory, may be presented as distinct kinds of structures. For instance, one may present it as a pair $\langle G, \circ \rangle$, with a non-empty set and a binary operation satisfying the axioms for groups. Alternatively, groups may be presented as $\langle G, \circ, e, - \rangle$, which we did before (this also satisfied the axioms for groups, which now involved the concepts of neutral element and opposite). Both kinds of structures are certainly distinct, and classes of the first kind are *not identical* to classes of the second. As Muller [Mul.11, sec.6] and Halvorson [Hal.12] argued, if we had an accompanying language, then we would be able to employ notions such as translatability or mutual definability to show both theories equivalent. But, given an implicit prohibition on the use of language due to the desire of being 'language-free', it seems that any such use would be troublesome. The general impression is that any appeal to language would throw us back to the Received View, which clearly is not the case! On the other hand, it is also argued that only language could help us in avoiding the problems, which is also clearly not the case. This illustrates how some cherished features of the semantic approach may also easily distract us from what is relevant.

Language was also recently involved in another controversy related to the concepts of morphisms — in particular, certain kinds of embeddings and isomorphisms. Isomorphisms, for instance, are important in the semantic approach, in particular for the relation between models of distinct theories, when we wish to claim that a theory is an extension of the other (horizontal relations between theories). They are also important to account for the relations between a model of the theory and data models (vertical relations), which are then responsible for the relation between a theory and experience. This is a main feature distinguishing the kind of semantic approach we are taking into account here from the Received View.

According to Suppes [Sups.60, Sups.67], for instance, and van Fraassen [vanF.80, p.56], a theory is applied to reality not directly, as in the case of the Received View, but through a hierarchy of structures. Roughly speaking (see [Sups.60] for details), we begin with simple experiments and construct a structure to model the phenomena. That is in general a qualitative model, which by itself does not take into account every feature of experience; as Suppes says, experience must be passed through a 'conceptual grinder' in order to be ready to be taken into account and related with a theory [Sups.67, p.62]. Later, it is necessary to find a numerical model isomorphic to the model of the phenomena. Only then can numbers be applied to things. This structure than may be embedded in still some larger structures before being embedded in a model of the theory. Models of the theory in general use mathematical concepts that don't have analogous in experience, and so cannot be directly applied to nature. So, even though this was a very rough description of the relation between theories and data, isomorphisms are involved throughout the whole process.

However, Halvorson [Hal.12] argued that the idea of isomorphism (and embeddings in general) do not make sense in the absence of a language. It is language that would be responsible for granting that each relation in a structure is mapped in a correspondent in another structure, which is the interpretation of the same symbol. In other words, without language, according to Halvorson, an isomorphism between structures $\langle D, R_i \rangle$ and $\langle E, K_i \rangle$ must be composed by both a bijection f between D and E and also by a function g sending each R_i into some K_j. An isomorphism then is such that for any n-ary R_i, we have

$$\langle d_1, \ldots d_n \rangle \in R_i \text{ iff } \langle f(d_1) \ldots f(d_n) \rangle \in g(R_i).$$

So, for instance, given a structure $\langle D, A_1, A_2 \ldots A_n \rangle$ and another structure $\langle E, B_1, B_2 \ldots B_n \rangle$ there is no way to grant that A_1 will correspond to B_1; that will happen only when $g(A_1) = B_1$, but that is not mandatory. The claim by Halvorson is that by having a language, the possible mixing of properties would be forbidden.

This definition of isomorphism is enough to allow for some tragedies, such as allowing that structures that differ in the true values they attribute to sentences turn out to be isomorphic (see [Hal.12, p.192] and [Lut.15] for a discussion). For instance, consider two theories in a first-order language with identity containing a denumerable family of unary predicate symbols P_1, P_2, ... as its non-logical vocabulary. Theory T_1 has as axioms only the formula $\exists! x(x = x)$, where $\exists!$ is the uniqueness quantifier; the axiom then states that there is only one thing in the domain of quantification. The interpretation of each P_i is then arbitrary on the domain containing only one element. Let T_2 be composed of the axiom of T_1 along with the infinite collection of formulas of the form $P_1 \rightarrow P_n$, for every n. Any model of T_2 will also have only one element in the domain of quantification, but will have to make sentences of the form $P_1 \rightarrow P_n$ true. Both theories are intuitively not equivalent, but, with the aforementioned notion of isomorphism, it is easy to show that for any model of T_1 there is an isomorphic model of T_2 and vice-versa

The main problem comes from the desire to make theories language independent. The fact that we cannot appeal to language to grant that, for instance, given a bijection f between the domains of interpretation, f must preserve the interpretation of P_1, allowing us to mix the interpretations of the predicate symbols (the role of the function g in the aforementioned definition). So, if the theories are isomorphic, they should be elementarily equivalent, which is clearly not the case.[2] Then being language-free poses some problems, according to Halvorson, which point to an inadequacy of the semantic approach on the individuation of theories.

As we shall see later, the notion of isomorphism employed by Halvorson is not the correct one. There is a standard notion of isomorphism between structures that employs the labels of the relations in a structure to grant that structures have a similarity kind. So only structures of the same similarity kind will allow for the definition of isomorphism without the need of a language. What is relevant is that structures have the form $\langle D, R_i \rangle_{i \in I}$, where the index set I orders the relations; isomorphisms must somehow observe that order: relations with the same indexes must be preserved. So isomorphisms may be defined both in the presence as well as in the absence of a language in which the structures are interpreted; both approaches to models as we have presented them are allowed to use the concept meaningfully.

So, perhaps, the conflict between structures that are interpreted in a formal language and structures that are not interpreted is not a substantial debate. In fact, that is precisely the conclusion arrived at by Lutz [Lut.15]: the debate between allowing a language and not allowing a language may be easily dispelled when we notice that a structure such as $\langle D, R_i \rangle_{i \in I}$, called *indexed structure*, may be converted into a Tarski-style structure when we allow the set I of labels to be the non-logical vocabulary of a language. If that is the case, the indexing is precisely an interpretation, and we have a *labeled structure*. There is no reason for accepting one of those kinds of structures and denying the other.

Our approach in this book will be to accept both kinds of structures (given that they are convertible into one another). This is in fact a lesson to be learned from Suppes's [Sups.67, p.60] presentation of the first features of the semantic approach. According to Suppes, a theory may be presented in two complementary ways: either following an *intrinsic approach* or else following an *extrinsic approach*. The intrinsic approach comprises the standard axiomatization using linguistic resources. When no such axiomatization is available, or is just too cumbersome, we may adopt the extrinsic approach of characterizing the class of models directly in set theory. For that, as we shall argue, we may employ what is called a 'Suppes predicate', which helps us in collecting the relevant models.

In the end, the distinction comes not from the kind of structure employed, but rather on styles of axiomatization. While employing formal languages, we think immediately about formal theories, with their axioms framed in the formal language when we talk about structures as purely set-theoretical constructs we are employing a Suppes predicate to gather the structures. Both approaches are interesting and each has its own advantages, as we shall see. So Suppes argued that the model approach is superior when it comes to studying the relation between theory and data and when axiomatization issues appear. In particular, he argued that using set theory as the language in which to develop the class of models allows us to avoid axiomatization of the required mathematics of a theory: all of the mathematics may be assumed as already developed inside set theory. So we may proceed directly to the empirical part of the theory that concerns us. Anyway, what is more relevant is that Suppes himself did not require that we abandon language in any sense (see in particular the remarks in [Sups.11, sec.2]). Philosophers of science may benefit from both approaches: they are representations of theories, and the most adequate approach will depend on what we need.

This is a lesson that should be already learned from Suppes's distinction between intrinsic and extrinsic characterization. That distinction was employed by da Costa and French [CosFre.03] for many purposes. In particular, the extrinsic approach (along with the partial structures approach) was employed to represent theories and their relation with experience. The intrinsic approach, on the other hand, was employed to discuss propositional attitudes such as what does it mean to believe in a theory. Belief is directed toward propositions, so it requires a linguistic approach. It is curious that the benefits of such a distinction went largely unnoticed. What is relevant for us is that theories are merely represented in the philosophy of science for philosophical purposes. What they really are goes beyond such representations, possibly. Our focus will be on the fact that semantic and syntactical approaches are representational tools for philosophical purposes.

The adherents of the semantic approach, then, went wrong in simply claiming that a theory may be identified with a class of models. That left them with no option but to employ identity of classes as identity of theories, which lead them astray. Also, that identification left the impression that a theory is independent of language. Again, much trouble appeared for the identity of theories. Difficult questions such as whether theories are the same or not troubled the view. We shall propose, then, in the next section, a more flexible consideration of both semantic as well as syntactic approaches, following the lines of Lutz [Lut.15], as well as [KraAreMor.11].

1.3   SCIENTIFIC THEORIES AND PHILOSOPHICAL TOOLS

Perhaps the greatest lesson we have learned from those discussions is simple: deep problems arise when we attempt to *identify* a theory with something (i.e., to reify theories, to use van Fraassen's [vanF.89, p.222] famous terms). The Received View faces many obvious inadequacies, for instance, if we identify a theory with a formulation comprising a specific language and a specific set of correspondence rules. In the same way, the semantic approach faces many difficulties if a theory is identified with a class of models: the corresponding criterion of theory identity will identify theories that are distinct and differentiate theories that are intuitively the same.

Facing all those difficulties, it seems that a distinct approach should be taken. Perhaps a good try would be to consider that both the syntactic approach as well as the semantic approach are not philosophical theories about the nature of something, *viz.*, scientific theories. Or even better, they should not be treated like that, even though they have been treated this way. As Lutz [Lut.15] has remarked, the whole debate was based on misunderstandings, and whether one decides to focus on languages or on set-theoretical structures depends on the main interests in the moment and on which approach affords more convenience. Philosophical approaches to scientific theories should not be censored for not capturing the *essence* of scientific theorizing; they strive at capturing aspects of such a diverse theorizing (see also the conciliatory note by Halvorson [Hal.15, p.15], according to whom "these approaches need not be seen as competitors").

This idea may be traced back to some of the first proponents of each view. Recall our brief discussion on Carnap on explication: as we mentioned, the goal of the Received View was not to provide for a general explication of the concept 'theory', but to provide general guidelines to study philosophical problems of scientific theories. Obviously, the problems and the guidelines were seen through the lenses of logical empiricism, but that is not the problem. What is relevant is that for many purposes of philosophical investigation, an approach is available that helps philosophers shape their questions and look for answers.

Again, as Lutz [Lut.12, sec.5] remarked, an explication could work as a rational reconstruction of an intuitive concept. In the case of particular theories, the Received View worked as an ideal form that scientific theories should be put in and from which philosophers could then study important relations between theoretical terms and their meanings and whatever other philosophical problem should interest them. Obviously, the Received View failed in providing a convincing framework in its overall form, but that happened for reasons that are generally not pointed out by the critics.

The same may be said of the semantic approach. In discussing the relation between theory and experiment, Suppes [Sups.67, p.63] noted that "there is no simple answer to be given" to the question "What is a scientific theory?" He further remarked that a precise answer to that question is not important. What is relevant is

> to recognize that the existence of a hierarchy of theories arising form the methodology of experimentation for testing the fundamental theory is an essential ingredient of any sophisticated scientific discipline.
> [Sups.67, p.64]

That is, for the point on which he was concerned, which was to understand the relation between theory and experiment, it matters more that we understand the proper relations of experience of the theory with testing than that we answer the question, what is a scientific theory? An for those concerns, he argued, an extrinsic approach is more appropriate. This is also clear from Suppes's [Sups.11, sec.2] observation that his approach is not the same as van Fraassen's: the latter attempts to formulate a theory as free of language and has left it very far from experiment. This is clearly not adequate for the purposes Suppes had in the moment.

So, taken seriously, those views by Carnap and Suppes seem to lead us to a distinct direction, other than those that the whole debate has taken. They seem to concede that the purpose of offering an approach to theories is not to capture the essence of a general concept, but rather to provide the formal tools for us to study and develop what for us, as philosophers, matters the most (which in general includes topics such as truth, belief in theories, acceptance, empirical adequacy, and so on). That is also the general line we shall take in the following chapters. They should not be read as an attempt to provide the tools for a *definition* of theories, but rather as furnishing frameworks in which certain kinds of studies may be taken. Our focus shall be on logical and metamathematical issues with foundational purposes, so this goal should also be kept in mind.

Perhaps it would be appropriate to insist again on a topic recently raised by French [Fre.15, p.14]: the semantic approach, the syntactic approach, and any other approach to theories are just *representations of theories* developed for philosophical purposes. They are not to be confused with the thing itself. What are those theories in and of themselves? Do theories even exist? Those are surely interesting issues, but they may be developed independently from the semantic and syntactical approaches. They are really to be treated as rational reconstructions and may be judged accordingly.

Having that point of view stated clearly, we shall now proceed to the topic of axiomatization of theories, perhaps the first step for foundational and metamathematical studies (which is what concerns us here).

NOTES

1. As we shall see later, we distinguish between the order of a language from the order of a structure, hence the distinction in the notation.
2. We recall that elementary equivalent theories are those first-order theories that satisfy the same first-order sentences. A precise definition can be seen, for instance, in [Men.97].

2 Axiomatization

Why to axiomatize? When we think about the logical foundations of science, the use of the axiomatic method arises as a rather natural idea. The axiomatic method is unique in allowing us to organize a field of knowledge and put it in a logically coherent structure; it clarifies and provides a sharp understanding of how distinct parts of the body of knowledge, when axiomatized, connect themselves, giving us the edifice of the whole discipline. Besides, the possibility of metatheoretical studies is another great advantage of rigorous formulation delivered by the axiomatic method, an advantage that must be explored for foundational studies. So, in the face of it, we take it that the axiomatic method is the preferred tool for the kind of study we wish to approach here. As we discussed rather briefly in the previous chapter, the main differences between approaches to theories may also be expressed as differences in styles of axiomatization. So a first look at the general features of the method is not only convenient, but also required.

Even though the use of the axiomatic method (AM) is nowadays common practice in mathematics, the historical roots of its use go back to the early Greeks and their study of philosophy. Of course, a great deal of applied mathematical knowledge had already been achieved by the Babylonians and Egyptians, but it stood more for 'collections of prescriptions' for doing practical calculations than for an organized field of knowledge [Sza.64]. These people did not seem to think of their mathematical knowledge as an edifice, resting on a set of basic accepted (i.e., not proved) propositions whose truth would be enough to derive the truth of all other known mathematical propositions. As Szabó suggests, that was achieved by the Greeks, and not in mathematics properly, but in philosophy:

> [they] [the Greeks] seem to have come to this [that is, to the idea that the starting principles should not be proved] from the practice of dialectic. They were accustomed to the fact that, when one of the partners in a debate wanted to prove something to the other, he was bound to start from an assertion accepted as true by both of them.
>
> [Sza.65][1]

An influential example suggested by this author was the use of indirect reasoning, now known as *reductio ad absurdum* by the Eleactics, mainly by Zeno of Elea.

Still in ancient times, Aristotle in his *Analytica Posteriora* declared that the axiomatic method was the method for the presentation of a deductive science (see [JonBet.10]). The basic idea behind the use of axiomatization is to present a certain field of knowledge in such a way that once certain assumptions are made, every other proposition that is true in the field can be deduced from these basic assumptions, today called *postulates* or *axioms*.[2] In this sense, the AM is to be applied to an already sufficiently known field of knowledge, and its main role would be 'hygienic', ordering propositions in the following sense: to avoid a regress to the infinite, we first select some set of concepts that we suppose are the basic notions to start with; they must be sufficient to allow us to define every other required concept. These basic notions are the *primitive notions* (or *primitive concepts*) of the system. The remaining concepts are *defined* from the primitive ones (also called *derived concepts*). The second step is to select some propositions, which are taken as *primitive*, consisting of the statements upon which the system is built. These propositions are written exclusively in terms of primitive notions and the defined concepts. All other statements of the theory must be obtained from the primitive ones by deduction and are then called *theorems* of the system. This kind of construction would make explicit the whole development of a field of knowledge: each concept employed is either primitive or explicitly defined, and each accepted proposition is either primitive or explicitly derived from previously accepted propositions.

As is well known, the paradigmatic example of an application of this method in ancient times is Euclid's *Elements* (4 BC), which attempted to systematize the mathematics known by the Greeks up to that time. Euclid's book has been a major influence in Western thought since then, an ideal of rigor to be pursued by mathematicians and philosophers until the nineteenth century. Its influence notwithstanding, it is now recognized that Euclid's presentation does not conform to the standards of rigor demanded by the modern use of this technique. On what concerns the presentation of primitive concepts, Euclid's book unexpectedly attempts to provide those terms with *definitions*. Some historians think those definitions could have been introduced not by Euclid himself, but by later commentators in order to 'explain' the meaning of the basic notions [Eucl.08]. As a result, it is hard to find in the *Elements* explicit primitive notions. On what concerns the theorems, which should be proved solely in terms of the axioms and previously proved theorems, the *Elements* also have some failures. For instance, the proof of Proposition 1 (Theorem 1) employs resources not covered by the previous assumptions (axioms or previously proved theorems).[3] Anyway, despite of the problems found in it, Euclid's book provided for a wonderful illustration of the general idea: every proposition to be accepted must be either postulated or else derived (deduced) from propositions which were postulated or already proven before.[4]

A new attitude on what concerns the axiomatic method was deeply involved with the development of mathematics itself. During the nineteenth century, mathematics went through another great transformation with the raise of abstract mathematical structures. Groups, rings, fields, geometries, algebras in general, topological structures, and all such rich varieties of abstract structures came to light and put this discipline in another paradigmatic level from which it will never return (but see the criticisms in chapter 4). The AM was essential for such developments. The nineteenth century also saw the birth of modern logic, which soon could also be studied as an abstract mathematical theory. The development of these two fields originated the mathematics of the twentieth century and perhaps still (in a sense to be discussed later in this chapter) of the twenty-first century too.

Recall that in Aristotle's and Euclid's time, the choice of primitive concepts and primitive propositions was subject to a condition of intuitive clarity and intuitive truth, respectively. Given that primitive concepts were intuitively clear, the defined concepts could be intuitively understood as well. Given that the primitive propositions were necessarily true and intuitively evident, the propositions derived from them were all true as well.[5] As a result of the revolution that began in the nineteenth century, the choice of the primitive notions and postulates are no longer subject to such demands. The postulates are assumed because they are useful for the purpose in mind: they *serve* for the intended finalities, namely, to derive the theorems. David Hilbert's 1899 axiomatization of Euclidean geometry makes use of three primitive notions, namely, those of *point, straight line*, and *plane*, and three relations holding among them, *incidence, betweenness* and *congruence*. Defined notions are, for instance, *right angle, parallel lines*, and so on (the definitions are not relevant for our purposes). But the Italian mathematician Mario Pieri presented an axiomatic for Euclidean and Bolyai-Lobachewskian geometry based on only two primitive notions: point and motion;[6] in this axiomatics, straight line and plane are *defined concepts*. From these axioms, all geometric theorems can be obtained. This shows that, in principle, there is no mathematical concept (and we could say the same also for the empirical theories) that cannot be defined; definability is relative to the language employed (see chapter 5 on definability).

Other assumptions from ancient times that are also relaxed include the demand that the postulates must be all *independent*; that is, a postulate should not be deducible from the remaining ones. This is taken now as a matter of convenience. Sometimes a *redundant* axiomatics is preferable for ease of treatment. A typical example is the Zermelo — Fraenkel set theory to be seen later, where the Pair Axiom may be derived from the Substitution Axiom and the Power Set Axiom. Historically, as the axiomatic method evolved, some of the ancient demands of intuitiveness were being left behind, with a crescent search for rigor taking its place.

So the axiomatic method not only benefited from the development of mathematics, but it also contributed to it. A new understanding of the method lies at the roots of the rise in abstraction in mathematics from the nineteenth century on. This 'paradigm shift' even gave rise to a famous dispute between Frege and Hilbert on the nature of the axiomatic method and its proper understanding (see [Bla.14]).

But leaving disputes aside, the success of the axiomatic method was something to be explored further, and this was precisely what Hilbert proposed in the turn of the twentieth century: the axiomatization of the theories of physics, which had been carried out more or less by chance, as in the case of Newton's physics—which can be regarded as a kind of axiomatics —,[7] was seen to be the next natural step of application of the method going beyond mathematics. This proposal was famously advanced by David Hilbert as the sixth of his celebrated list of 23 Problems of Mathematics, presented at the II International Congress of Mathematicians in 1900. There, Hilbert stressed that

> whenever, from the side of the theory of knowledge or in geometry, or from the theories of natural or physical science, mathematical ideas come up, the problem arises for mathematical science to investigate the principles underlying these ideas and so to establish them upon a simple and complete system of axioms, that the exactness of the new ideas and their applicability to deduction shall be in no respect inferior to those of the old arithmetic concepts.
>
> [Hilb.76, p.5]

And, later,

> [t]he investigation on the foundations of geometry suggest the problem: *To treat in the same manner, by means of axioms, those physical sciences in which mathematics plays an important part; in the first rank are the theory of probability and mechanics.*
>
> (ibid., p.14)

His ideas on this respect were already expressed some years before; in 1894, after discussing the role of the AM in geometry, he stressed that

> [n]ow also all other sciences are to be treated following the model of geometry, first of all mechanics, but then also optics and electricity theory.
>
> (quoted from [Sau.98])

We will give examples of axiomatics in the sciences in chapter 5, but here we just mention that during the twentieth century, much was done in the directions pointed out by Hilbert not only in physics but also in other disciplines, such as biology, initiated by John Woodger (See [All.38] but with several developments after him. See also [Will.70], [Jong.85], [MagKra.01], [Esan.13], where further references can be found. For a general setting in the beginnings of much of these applications, see the papers presented in [Hen.et al.59].) As for the application of the AM in psychology, see [Bog.79] for an account of Suppes's works on the subject. But the most investigated (from the axiomatic point of view) disciplines were those of physics, such as classical particle mechanics [Sups.02], continuum mechanics (see [Ign.96]), thermodynamics.

Now, for a sharper illustration of the advance in the direction of crescent rigor and abstraction on the development of the AM, we shall distinguish among three *levels of axiomatization*. We notice in advance that this classification is not the only one possible; Hilbert distinguished between *concrete axiomatics* and *formal axiomatics* [Hil.96]; Jean Ladrière went deeper by distinguishing among *intuitive axiomatics*, *abstract axiomatics*, *formal axiomatics*, and *pure formal systems* [Lad.57, pp.36ff], but we think that our three levels capture the distinctions we need. Our distinction will be as follows: *intuitive* (or *concrete*) *axiomatics*, *abstract axiomatics*, and *formal axiomatics*. Their meanings and examples are given in the next sections.

2.1   CONCRETE AXIOMATICS

In concrete axiomatic systems, which illustrate the way the Greeks used the AM, the scientist has a well-defined field of knowledge or domain of application in mind: the axiomatics is meant to reflect the logical structure of a (supposedly) well-known domain. The basic concepts of the axiomatics are already interpreted in that domain; the basic propositions are true of that domain. Then the axiomatization aims at the organization of the field by presenting in an organized fashion its basic concepts, basic assumptions and all the possible results that can be derived from this basis. In his paper "Axiomatic Thought", Hilbert put things this way:

> When we assemble the facts of a definite, more-or-less comprehensive field of knowledge, we soon notice that these facts are capable of being ordered. This ordering always comes about with the help of a certain *framework of concepts* [*Fachwerk von Begriffen*] in the following way: a concept of this framework corresponds to each individual object of the field of knowledge, and a logical relation between concepts corresponds to every fact within the field of knowledge. The framework of concepts is nothing other than the theory of the field of knowledge.
>
> [Hil.96, pp.1107–8]

That is, by organizing a certain field of knowledge known in advance, we enlighten its 'frame of concepts', which is, of course, a free choice of ours (recall that Pieri had chosen an unusual class of primitive notions than those used in most formulations of Euclidean geometry). In this sense, axiomatization would not have much to do with the *development* of the framework, strictly speaking, except for the new theorems that can be proved (but for some critics, as we shall see later, these theorems are 'already implicit' in the postulates, so nothing new would be introduced in the theory); it just *cleans the house*. As we shall see later, this is a misconception.

Besides Euclid's axiomatics of geometry, typical examples of concrete axiomatization include arithmetics (in the sense we will discuss in a few moments), which deals with (natural) numbers and their operations, and Zermelo's set theory, which axiomatizes the concept of set (collections of objects) and objects that can be members of the sets. But let us begin with a simple case, that of Patrick Suppes's theory of human paternity just to emphasize the technique [Sup.77]. Suppes had in mind a specific domain of application, namely, human beings. His theory does not apply to either worms or plants and can be summarized as follows. The primitive notions are *alive human being*, *male human being*, and *father of*, which apply to two human beings (other choices are possible, but here we follow Suppes).

The postulates are

(P1) If the human being *a* is father of the human being *b*, then it is not the case that *b* is father of *a*.
(P2) All living human beings have only one father who is a male human being.
(P3) All living human beings have only one father who is not a male human being.

The human being introduced by (P3) can be defined as *mother* of the given human being. Other defined concepts are *grandfather*, *brother*, *sister*, etc. Some theorems can be easily stated and proved: the father and the mother of a given human being are different human beings; no human being is the father of him or herself, and so on. It is important to notice that, although the axiomatics can have other *models*, in a sense to be made clear later, the *intended domain* is already given in advance. In elaborating this theory, we *already know* (by hypothesis) the main traits of human paternity theory.

Intuitive Peano's arithmetic is our second example. The domain of application is composed by the natural numbers, which for us includes zero (Peano started with 1). Supposedly, we know already what to do with natural numbers, their main characteristics, and operations. Then we may take as primitive notions the following: *zero* and *sucessor* (of a natural number). Defined concepts are, for instance, *one*, *two*, *three*, *prime number*, *even number*, and so on. The typical postulates are:

(PA1) Zero is not a successor. In other words, there is no natural number from which zero is a sucessor.

(PA2) If two natural numbers have the same successor, they are the same natural number.

(PA3) If a certain property applies to natural numbers, then if zero has this property and whenever a natural number has the property it follows that its successor has it also, then all natural numbers have that property.

This theory may also have other *models* (realizations of the axioms; the concept of model is being used here informally, but will be considered later); for instance, think of the same sequence of the (intuitive) natural numbers, but now call 'zero' the number 100 and, as a sucessor of a certain number *n*, the number *n* + 100. Then it is clear that the axioms are satisfied, and all other concepts and operations can be adapted to this case. This would be a strange arithmetic to be used in a drugstore, but from the mathematical point of view that doesn't matter. The system, as an interpreted calculus, *works*! As we have seen, the nature of the *models* of a theory is something to be carefully analyzed.

A third example of a concrete axiomatics is Zermelo's set theory presented in 1908 [Zer.67]. Zermelo provided an axiomatic system for Cantor's theory of sets, seen as collections of objects. The very notion of set was characterized by Cantor himself as follows:

> by an "aggregate" (*Menge*) we are to understand any collection into a whole (*Zusammenfassung zu einein Ganzen*) M of *definite and separate objects m* of our intuition or our thought.
>
> (our emphasis [Low.14, p.85])

In other words, a set is a collection of objects of whatever sort with the proviso that their elements are distinct from one another.[8] We can form sets of human beings, ants, angels, hurricanes, gods, prime numbers. But of course Cantor was thinking (again!) of mathematics.[9] However, as is well known, the superb theory he created was inconsistent. The so-called paradoxes of set theory clearly show this in a unequivocal way (a very clear discussion can be found in [FraBarLev.73]).

To axiomatize set theory, Zermelo said that "[s]et theory is concerned with a *domain* 𝔅 of individuals, which we shall call simply *objects*, and among which are the *sets*" [Zer.67]. Those individuals that are not sets are the *atoms* (*Urelemente* in the German terminology). The fundamental relation is that one which says a certain individual *a* is an element or belongs to a set *b*; here we shall write $a \in b$ to use an updated notation. This is the membership relation. As we shall see from the axiomatics, the atoms may be elements of sets; hence, sets may have as elements either atoms or other sets. Atoms do not have elements. Zermelo still introduces the notion of *subset*: a set *x* is a subset of a set *y* if every element of *x* is an element of *y*. In our updated terminology, we write $x \subseteq y$ to express that. Needless to say, we are not using Zermelo's original terminology.

The aim of Zermelo's axiomatization was to avoid some inconsistencies resulting from the far too liberal notion of set used in Cantor's informal theory. However, in stating his postulates, Zermelo also did not fulfill all the logical details. For instance, in his Separation Axiom (discussed later in this chapter), the basis of his development, he used the rather vague notion of a *definite property* in order to form sets from already given sets by 'separating' in the given set those elements that fulfill the given property. It took some time until Skolem and Fraenkel proposed to replace the notion of a 'definite property' with a precise characterization, as we shall see later in this chapter.

The postulates are as follows:

[Axiom of Extensionality] If two sets have the same elements, then they are the same set. In other words, given two sets x and y, if every element of x is an element of y and vice-versa then $x = y$.

This axiom makes it useless for a set to have repeat elements. If we write {1, 2, 3} for the set comprising 1, 2, and 3 as its elements, then by force of this axiom, it is identical to the set represented by {1, 1, 2, 3, 3, 3}, for they have the same elements.

[Axiom of the Elementary Sets] There exists a ('fictitious', according to Zermelo) set, the *null* set that contains no element at all. Given an object a, either a set or an ur-element, there exists a set {a} containing just a as its element. Given objects a and b, there exists a set {a, b} whose only elements are a and b and nothing else.

The set {a} is called the *unitary* of a and can be derived from {a, b} when $a = b$. Hence it suffices to postulate the existence of the *unordered pair* {a, b}. Uniqueness of the null set (or 'empty set' in modern language) can be proved and it is denoted today by $\emptyset$.

[Axiom (Scheme) of Separation] Whenever the formula $F(x)$ (Zermelo spoke in terms of 'propositional function') is *definite* for all elements of a set x, then there is a subset y of x formed by precisely those elements of x that fulfill the condition $F(x)$.

As we have already stated, the notion of 'definite property' was not satisfactorily clear. Today it is common to axiomatize set theory as a first-order system, as we shall see in the next chapter. Then $F(x)$ is a formula of a first-order language with just one free variable, x. But, intuitively, the result is the same: given a set x and a condition F, we can 'separate' from x those elements that fulfill the condition and with them form another set y. This trick is essential for avoiding the existence of *enormous* sets, which originate paradoxes in Cantor's theory, for instance, the set of all sets or the set of all those sets not belonging to themselves. So the whole domain $\mathfrak{B}$

of sets and the *Urelemente* is also not a set.[10] Zermelo's axiom is really a *scheme*, for it potentially originates an infinity of axioms, one for each formula *F* we use.

The notion of *subset* was mentioned already: a set *x* is a subset of a set *y* if all elements of *x* are also elements of *y*. In today's terminology, we write $x \subseteq y$. Some basic properties are immediately obtained: (a) the empty set is subset of any set, (b) every set is a subset of itself, (c) if $x \subseteq y$ and $y \subseteq x$, then $x = y$ and so on.

[Axiom of the Power Set] To every set *x* there corresponds a set $\mathcal{P}(x)$, the *power set* of *x*, whose elements are the subsets of *x*.

[Axiom of Union] To every set *x* there corresponds a set termed $\bigcup(x)$ whose elements are the elements of the elements of *x*.

If the set *x* has just two sets *a* and *b* as elements, then it is common today to write $a \cup b$ to indicate the union of *x*.

[Axiom of Choice] If *x* is a set whose elements are also non-empty sets and pairwise disjoint sets, then there exists a set *y* having one and only one element in common with each element of *x*.

[Axiom of Infinity] There exists a set that contains the empty set and, having *a* as an element, has also {*a*} as an element.

The natural numbers (Zermelo's natural numbers) are defined (defined concepts) as follows: $0 := \emptyset$, $1 := \{0\}$, $2 := \{\{0\}\} = \{1\}$, $3 := \{\{\{0\}\}\} = \{\{1\}\} = \{2\}$, etc. All set theory, including Cantor's theories of transfinite ordinals and cardinals, functions, orders, and so on can be developed from this axiomatic basis.

Again, we have here an axiomatics of a certain field known in advance, namely, set theory. The theory just selects some collections to be sets so that the known paradoxes do not arise.[11]

Axiomatizations in the empirical sciences are in general concrete axiomatics, for we are always thinking of an already known field of knowledge, for instance Darwinian selection theory (see [Will.70]) or classical particle mechanics [Sups.02], yet the resulting theories may have other 'models', as we shall see later.

It is important to realize that different axiomatizations of the same domain are possible. Depending on the chosen primitive notions and axioms we use, different versions of the theories can be achieved. The discussion of whether two different axiomatizations lead us to different theories was central in the traditional debate about the nature of scientific theories (see again our chapter 1); here we assume that two different versions of group theory, say by choosing different primitive concepts or axioms, are versions of the same theory. In the next section, we will comment on this point a little bit more.

2.2   ABSTRACT AXIOMATICS

A further step in the process of rigor in axiomatization appears with the rise of abstract axiomatics. The reader should note that the three levels of axiomatic theories we are considering *are different levels of abstraction*. Abstract axiomatic systems don't have a fixed domain of discourse, although in general there is a domain which motivates the development of the axiomatics. So abstraction from a domain of application does not entail that it has no *preferred* or *intended* domain. On the contrary, in general, abstract axiomatics arise from the study of a certain field; in the sense of concrete axiomatics, when the scientist realizes that it can have other *models*, or domains of application, she develops her theory in such a way that she may cope with all these domains at once. In this sense, the axiomatic method is said to provide for an economy in thought.

In the paper The architecture of mathematics [Bou.50], Bourbaki claims that the rise of abstract axiomatics characterizes modern mathematics, for the AM has the capacity of unifying fields that could, in principle, be seen as distinct, so avoiding a 'tower of Babel' of disciplines. The infinitely many groups, for instance, the infinitely many linear (or vector) spaces, the infinitely many topological spaces, and so on can now be studied within a unified framework, one for each category of mathematical object (groups, linear spaces, topological spaces, etc.).

Typical abstract axiomatic theories are those mentioned earlier as well as algebraic rings, fields, and so on. Here the structural aspect of mathematics appears for the first time in a clear way. For instance, group theory can be characterized as the mathematical study of structures of the following kind:[12] there is a non-empty domain G and a binary operation $\circ$ on G such that this operation is associative, admits a neutral element e, and every element $a \in G$ has an inverse $-a \in G$ relative to the operation, as we shall see in more detail later in this chapter. Historical details about groups can be easily found in the literature, going back to the studies of zeros of certain polinomial equations. But the very concept of group soon became abstract, in the sense that neither the domain G nor the binary operation are specified, except for a particular application. Groups are *abstract mathematical structures* of the form $\mathcal{G} = \langle G, \circ \rangle$. We may find infinitely many structures that fulfill the definition, that is, cases of *groups*. The theory of groups studies the mathematical properties of these entities also in an abstract way. Also, group theory can be applied to several domains beyond mathematics, such as physics; Hermann Weyl has pioneered the use of groups in quantum mechanics.

Just to make justice to what was said in the end of the last section about different ways of axiomatizing a same field of knowledge, let us mention that we can formulate the abstract notion of a group as follows: we take a structure $\mathcal{G} = \langle G, \circ, e, - \rangle$ comprising a non-empty set G and a binary operation $\circ$ defined on G as before, but now we add other primitive elements to the structure, say a designated element e of G and an unary function $-$

from G to G to form the inverse elements of the elements of G. Now it is enough to adequately modify the axioms given. But the *theory*, here understood as comprising all the results that can be obtained from these axiomatics, is the same as before. Recall the thorny problems faced by the semantic and syntactic approaches to theories when no such equivalence between formulations is allowed.

It is important to emphasize that in this kind of axiomatization, we do not have a particular a priori intended domain of application. Linear spaces, despite their motivations, have also infinitely many instantiations, or *models*. For instance, Hilbert spaces, which are important in quantum mechanics, are linear (vector) spaces of a kind.

Concrete axiomatic systems can be transformed in abstract systems once we 'abstract' the primitive notions of their intuitive meaning. This is a natural move to abstraction and is precisely what the history of the method tells us about its evolution. The rise of abstract mathematical structures was achieved precisely when mathematicians noticed that their concrete axiomatics could be applied to other domains beyond the intended one.

In abstract axiomatic systems, however, something is still kept in the shadows implicitly assumed: the underlying logic. By *logic*, except when explicitly mentioned, we are being liberal so that it involves either higher-order logics or set theory (we shall leave category theory out of this discussion). Thus in presenting group theory, we are neither making explicit the mechanisms we use to make deductions nor explaining that the binary operation ○ is a mapping from the cartesian product $G \times G$ in G, that is, a set. The different ways of axiomatizing and their relations to traditional philosophical approaches to scientific theories will be discussed in a later chapter.

Our second example of abstract axiomatics is that of a Hilbert space, the kind of theory we will need when considering quantum mechanics.[13] In short, a Hilbert space is a linear vector space with an inner product that is complete in the norm defined by the inner product. Let us be more specific. A linear (or vector) space is a structure of the form $\mathcal{E} = \langle \mathcal{V}, \mathcal{K}, +, \cdot \rangle$, where

(1) $\mathcal{V}$ is a non-empty set whose elements are called *vectors*. This name has its origins in the intended model of linear spaces, namely, that where the vectors are geometric vectors (sometimes also called Euclidean vectors — that is, quantities having a length and a direction, frequently used in physics and in vector algebra). We shall use small Greek letters such as α, β, ... for denoting vectors.

(2) $\mathcal{K}$ is a *field*, which by itself comprises other elements; $\mathcal{K}$ is by itself a mathematical structure $\mathcal{K} = \langle K, +_K, \cdot_K \rangle$ obeying the postulates of fields. In most applications to physics, $\mathcal{K}$ is taken to be either the field of the real numbers or the field of complex numbers, with the last one being the most relevant for quantum mechanics. The elements of the domain K of the field are called *scalars* and denoted by small Latin letters a, b, x, y, ..., sometimes with indexes.

(3) $+$ is an application[14] from $V \times V$ to V, called *vector addition*. The image of the pair $\langle \alpha, \beta \rangle$ is written $\alpha + \beta$ and called *the sum* of the vectors α and β. It is postulated that $\langle V, + \rangle$ is a commutative group whose neutral element is called *the null vector*, denoted by **O**. The inverse of α in this group is written $-\alpha$.

(4) $\cdot$ is an application from $K \times V$ in V, called *product of a vector by a scalar*. The image of the pair $\langle x, \alpha \rangle$ is written $x \cdot \alpha$ or just $x\alpha$. In physics, usually this is also written $\alpha \cdot x$ or just αx by an abuse of language. It is postulated that this operation satisfies the following axioms, for any $\alpha, \beta \in V$, and $x, y \in K$:

(a) $x(\alpha + \beta) = x\alpha + x\beta$

(b) $(x +_K y)\alpha = x\alpha + y\alpha$

(c) $(x \cdot_K y)\alpha = x(y\alpha)$

(d) $1\alpha = \alpha$, where 1 is the neutral multiplicative element of the field K

Later we shall speak of particular linear spaces (or of 'models' of this structure). We shall proceed here as does the standard mathematician. By an abuse of language, we follow the standard terminology and say that W is a linear (or a vector) space over the field K. When $K = \mathbb{R}$ — that is, the field is the field of real numbers — we speak of *real* linear spaces and in *complex* linear spaces when $K = \mathbb{C}$.

Definition 2.2.1 (Restriction of an operation) *Let X be a set and $\circ$ a binary operation on X. Denoting as usual by $a \circ b$ the image of the pair $\langle a, b \rangle \in X \times X$ by the mapping $\circ$, let $Y \subseteq X$. We call the restriction of the operation $\circ$ to Y, denoted $\circ_Y$ the operation defined by $a \circ_Y b$ iff $a \circ b$ and $a, b \in Y$.*

In other words, it is 'the same' operation but now considered only in relation to the elements of the subset Y. It is easy to extend this definition to most general cases as those used in the next definition.

Definition 2.2.2 (Subspace) *A structure $\mathcal{E}' = \langle W, K, +_W, \cdot_W \rangle$ is a subspace of the linear space $\mathcal{E} = \langle V, K, +, \cdot \rangle$ if $W \subseteq V$ and $+_W$ and $\cdot_W$ are the restrictions of the operations of $\mathcal{E}$ to the sets in $\mathcal{E}'$, they are also a linear space over the same field K.*

The usual parlance in mathematics does not distinguish between the notations of the operations and their restrictions, leaving this distiction to the context.

In order to obtain a Hilbert space in which we may develop quantum mechanics, we fix a complex linear space (henceforth $K = \mathbb{C}$) and enlarge the structure with an additional operation called an *inner product*, which is a mapping now from $V \times V$ to $\mathbb{C}$. The image of the pair $\langle \alpha, \beta \rangle$ will be denoted by $\langle \alpha | \beta \rangle$. We call attention to the differences of notation: $\langle \alpha, \beta \rangle$ is an ordered pair of vectors, while $\langle \alpha | \beta \rangle$ is a scalar (a complex number).

The postulates to be satisfied by this new notion are the following ones, holding for all vectors and scalars:

(a) $\langle \alpha|\beta + \gamma \rangle = \langle \alpha|\beta \rangle +_K \langle \alpha|\gamma \rangle$
(b) $\langle \alpha|x\beta \rangle = x \cdot_K \langle \alpha|\beta \rangle$
(c) $\langle \alpha|\beta \rangle = \langle \beta|\alpha \rangle^*$, where the exponentiation * indicates the complex conjugation (remembering that is $z = a + bi$, then $z^* = a - bi$)
(d) $\langle \alpha|\alpha \rangle \geq 0$ and $\langle \alpha|\alpha \rangle = 0$ iff $\alpha = \mathbf{O}$

Then we introduce the following definition:

Definition 2.2.3 (NORM) *If $\alpha \in \mathcal{V}$, the norm induced by the inner product of the vector α, denoted $\|\alpha\|$, follows:*

$$\|\alpha\| := \sqrt{\langle \alpha|\alpha \rangle}.$$

It results that the norm of a vector is a real number. This norm is said to be *the norm induced by the inner product.*[15] With this notion, the concept of *distance* between two vectors α and β can be defined as follows:

Definition 2.2.4 (DISTANCE) *The distance between α and β, denoted $\mathsf{d}(\alpha, \beta)$, is given by*

$$\mathsf{d}(\alpha, \beta) := \| \alpha + - \beta \| = \| \alpha - \beta \|.$$

We just remark that, concerning the vector $-\beta$, the following identity can be easily checked: $-\beta = (-1)\beta$. Let $\alpha_1, \alpha_2, \ldots$ be a sequence of vectors in $\mathcal{V}$ (denoted by (α_i)), and let β be a vector in $\mathcal{V}$. Then

Definition 2.2.5 (CONVERGENCE) *We say that the sequence (α_i) converges to β iff for any real number $\varepsilon > 0$, there exists a natural number n such that for any $j > n$, $\mathsf{d}(\alpha_j, \beta) < \varepsilon$.*

Definition 2.2.6 (CAUCHY SEQUENCE) *A sequence (α_i) of vectors is a Cauchy sequence if for any real number $\varepsilon > 0$ there exists a natural number n such that for any $j, k > n$, we have that $\|\alpha_j - \alpha_k\| < \varepsilon$.*

Every Cauchy sequence converges, as it can be proved. The problem is that if we suppose a collection of vectors of $\mathcal{V}$ and assume that all vectors of the sequence are in this set, the vector to which it converges may not be in this set. In particular, this set of vectors, endowed with the restrictions of the linear space operations, may also be a linear space. In this case, we say that the corresponding structure is a *subspace* of the given linear space. Then we have the main definition:

Definition 2.2.7 (HILBERT SPACE) *A Hilbert space is a linear space with an inner product such that all Cauchy sequences of vectors converge to a vector still in the space.*

A subspace of V that obeys this condition is termed a *closed subspace*. The Hilbert spaces are the mathematical frameworks used in the most usual formulations of non-relativistic quantum mechanics (see the formulation of quantum mechanics in chapter 5).

Beyond the particular examples developed, which are useful for the discussions to come, the important thing at this point is to note that the mathematical structures being obtained (groups, linear spaces, Hilbert spaces) are abstract in the sense of our characterization. They don't have a particular domain being supposed in advance. But the formulations are still supposing the use of concepts such as mappings, sets, and so on, and the proofs (if we would provide some) assume an implicit underlying logic. So we need to go further.

2.3 FORMAL AXIOMATICS

Formal axiomatic theories go deeper in abstraction and rigor by making the underlying logic totally explicit and, in general, by using formal languages to express the propositions of the theory. Being explicit about the logic is really necessary, mainly because the logic employed could not be standard (classical) logic. The relevance of making logic explicit became clear during the last century when several deep metamathematical results where achieved. Gödel's incompleteness theorems, Tarski's theorem on the undefinability of the notion of truth, the existence of non-standard models of Peano's arithmetics, the independence proofs in set theory, and much more were possible only after the full understanding of the logic underlying classical mathematical theories was achieved. In particular, the rise of non-classical logics was possible only after logic achieved total formalization so that several other systems could be developed either by extending it ('complimentary logics') or by modifying some of its basic laws ('heterodox logics'). On what concerns empirical sciences, it is still a challenge to convincingly argue that physics, for instance, *really* needs a non-classical logic. We shall return to this point later.

For the moment, let us present here the main systems we shall be dealing with in this book as examples of formal axiomatics. The systems to be presented are all 'classical', that is, grounded on classical first-order logic, so let us present it first. By doing so, we keep the text self-contained.

2.3.1 First-Order Classical Logic With Identity

This logic, let us call it $\mathcal{L}^1$, will be presented through the 'linguistic' approach[16] by describing a basic vocabulary, rules of formation (the grammatical part of $\mathcal{L}^1$), and its postulates. Let us call the language L. The basic (primitive) vocabulary of L comprises the following categories of primitive symbols (as is common in these presentations, we do not distinguish between use and mention):

1. The propositional connectives: $\neg$ and $\rightarrow$.
2. A denumerable collection of individual variables: $x_0, x_1, x_2, \ldots$. We shall use $x, y, z, \ldots$ as metavariables for individual variables.
3. The universal quantifier: $\forall$.
4. Auxiliary symbols: left and right parentheses and the comma.
5. A collection (possibly empty) of individual constants: $a_1, a_2, \ldots$. We use $a, b, c \ldots$ as metavariables for individual constants.
6. For any natural number $n > 0$, a collection of predicate symbols of rank n: $P_1^n, P_2^n, \ldots$. We use $F, P, Q, \ldots$ as metavariables for predicate symbols, and the context will indicate their rank.
7. For any natural number $n > 0$, a collection of functional symbols (symbols for operations) of rank n: $f_1^n, f_2^n, \ldots$. We use $f, g, h, \ldots$ as metavariables for operations.
8. The identity symbol: $=$.

The symbols in 5, 6, and 7 are called *specific or non-logical symbols* and depend on the theory having $\mathcal{L}^1$ as underlying logic, as we shall see in the examples that follow. The other symbols are the *logical symbols* and can be taken as being the same for all theories.[17]

The grammar is described as follows: first, we define the *expressions* of L as sequences of finite symbols of L written horizontally from left to right. For instance, $\neg((x_1 x_2 \rightarrow A_4^3(\forall$ is an expression.

Definition 2.3.1 (TERMS OF L) *The terms of the language are the individual variables, the individual constants, and the expressions of the form $f(t_1, \ldots, t_n)$, where f is a functional symbol of rank n and $t_1, \ldots, t_n$ are n terms. These are the only terms.*

Definition 2.3.2 (FORMULAS) *The formulas of L (well-formed expressions) are defined by the following clauses: (1) If t_1 and t_2 are terms, then $t_1 = t_2$ is a formula called 'atomic'. (2) If F is a predicate symbol of rank n and $t_1, \ldots, t_n$ are terms, then $F(t_1, \ldots, t_n)$ is a formula (also 'atomic'). (3) If α and β are formulas, then expressions such as $(\neg\alpha)$ and $(\alpha \rightarrow \beta)$ are formulas. (4) If x is an individual variable and α is a formula, then $\forall x \alpha$ is a formula. These are the only formulas of L.*

In writing formulas, we make a standard convention for elimination of parentheses: for instance, we can eliminate the external parentheses in writing formulas. So $\neg\alpha$ abbreviates $(\neg\alpha)$ and $\alpha \rightarrow \beta$ abbreviates $(\alpha \rightarrow \beta)$. Parentheses are useful for avoiding ambiguities in writing formulas. Notice that α and β are also metavariables for formulas. Symbols like these (Greek lowercase letters) will be used; thus the expressions we shall call 'axioms' are in reality *axiom schemata*, which enable us to obtain axioms when uniform substitutions of formulas for these metavariables are performed.

Some additional syntactical definitions and conventions are in order to facilitate the exposition. An occurrence of a variable x in a formula is *bound* if x is the variable that appears just after the universal quantifier or if it occurs in a formula that is affected by (is in the *scope* of) the quantifier. Otherwise, it has a *free* occurrence in the formula. For instance, in the formula $\forall x(x = y) \rightarrow \forall y(x = y)$, the two first occurrences of x are bound and the third one is free, while the first occurrence of y is free and the last two are bound. If a formula α has $x_{i_1}, \ldots, x_{i_k}$ among its free variables, we write $\alpha(x_{i_1}, \ldots, x_{i_k})$. So $\alpha(x)$ stands for a formula that has free occurrences of x.

A term t is *free* for the variable x in the formula α if no free occurrence of x in the formula α lies within the scope of a quantifier $\forall y$, where y is a variable occurring in t. For instance, if f is a functional symbol of rank 2, then $f(x, y)$ is free for x in the formula $\forall z(x = z) \rightarrow (x = w)$, but is not free for x in $\neg\forall y\neg(x = z) \rightarrow (x = w)$.[18]

The postulates of $\mathcal{L}^1$ are

(L1) $\alpha \rightarrow (\beta \rightarrow \alpha)$
(L2) $(\alpha \rightarrow (\beta \rightarrow \gamma)) \rightarrow ((\alpha \rightarrow \beta) \rightarrow (\alpha \rightarrow \gamma))$
(L3) $(\neg\alpha \rightarrow \neg\beta) \rightarrow ((\neg\alpha \rightarrow \beta) \rightarrow \alpha)$
(L4) $\forall x\alpha(x) \rightarrow \alpha(t)$, where t is a term free for x in $\alpha(x)$
(L5) $\forall x(\alpha \rightarrow \beta) \rightarrow (\alpha \rightarrow \forall x\beta)$, provided that α does not contain free occurrences of x
(L6) $\forall x(x = x)$
(L7) $x = y \rightarrow (\alpha(x) \rightarrow \alpha(y))$, where $\alpha(y)$ arises from $\alpha(x)$ by replacing some free occurrences of x by y

Furthermore, our system comprises the following *inference rules*, where $\vdash$ is the symbol of deduction:

(MP) (Modus Ponens) $\alpha \rightarrow \beta, \alpha \vdash \beta$
(Gen) (Generalization) $\alpha \vdash \forall x\alpha$

The notion of deduction is crucial in characterizing a logic. If we modify it, we change the logic. The standard definition is the following one, where Γ denotes a set of formulas:

Definition 2.3.3 (DEDUCTION) *We say that a formula α is deduced, derived, and inferred from a set Γ of formulas (the premises of the deduction) if the following clauses are obeyed:*

(i) There is a sequence of formulas $\beta_1, \beta_2, \ldots, \beta_n$ such that
(ii) β_n is α

(iii) Each β_i $(i < n)$ of the sequence is either

(a) an axiom, or

(b) an element of Γ, or

(c) immediate consequence, by one of the rules of inference, of preceding formulas in the sequence

If these clauses are fulfilled, we write

$$\Gamma \vdash \alpha$$

There are situations in which $\Gamma = \emptyset$; that is, there are no premises in the deduction other than the axioms of the logic. In this case, we write

$$\vdash \alpha$$

and say that α is a *thesis* or a *formal theorem* of the logic.

Further useful definitions are

Definition 2.3.4 (OTHER CONNECTIVES AND THE EXISTENTIAL QUANTIFIER) *Taking into account our conventions for the use of parentheses, we put*

(i) $\alpha \wedge \beta := \neg(\alpha \rightarrow \neg\beta)$

(ii) $\alpha \vee \beta := \neg\alpha \rightarrow \beta$

(iii) $\alpha \leftrightarrow \beta := (\alpha \rightarrow \beta) \wedge (\beta \rightarrow \alpha)$

(iv) $\exists x\alpha := \neg\forall x\neg\alpha$

We get a *first-order theory* when we specify a particular choice of the so-called *non-logical symbols*, namely, individual constants, predicate symbols, and functional symbols and add further *specific axioms* (or *proper axioms*) for the particular theory. Thus the axioms (or postulates) of a first-order theory will be those of $\mathcal{L}^1$ plus those of the specific theory. In this sense, $\mathcal{L}^1$ is by itself a first-order theory, with no additional proper postulates.

For instance, we can formulate Suppes's theory of human paternity as a first-order theory as follows. We assume all the logical vocabulary given earlier of a first-order logic and consider the following specific primitive symbols: a binary predicate P and two unary predicates L and M. The only terms are the individual variables x, y, z, ... The atomic formulas are $x = y$ and the expressions of the form $P(x, y)$, $L(x)$, and $M(x)$. The intended interpretation of these formulas is obvious: x is a father of y, x is a living human being, and x is a male human being. The postulates, in addition to those given earlier, may be written as follows:

(P1) $\forall x\forall y(P(x, y) \rightarrow \neg P(y, x))$

(P2) $\forall x(L(x) \rightarrow \exists!y(M(y) \wedge P(y, x)))$

(P3) $\forall x(L(x) \rightarrow \exists!y(\neg M(y) \wedge P(y, x)))$

Here $\exists!x$ means 'there exists just one x'.

The next example of a first-order theory is Peano's arithmetics. We abbreviate $\neg(t_1 = t_2)$ by $t_1 \neq t_2$.

2.3.2 Peano Arithmetics

The underlying logical apparatus of the theory $\mathcal{PA}$ is the one presented earlier. We maintain the same logical symbols, but the specific symbols are the following: (1) an individual constant '0', (2) a unary functional symbol 'σ', and (3) two binary functional symbols $\oplus$ and $\otimes$. Hence, besides the individual variables, the terms of the language of $\mathcal{PA}$ are obtained from expressions such as 0, σt where t is a term, $t_1 \oplus t_2$ and $t_1 \otimes t_2$, for t_1 and t_2 any terms.

The postulates of $\mathcal{PA}$ are (L1)–(L7) plus the following:

(PA1) $\forall x(0 \neq \sigma x)$
(PA2) $\forall x \forall y(\sigma x = \sigma y \rightarrow x = y)$
(PA3) If $\alpha(x)$ is a formula with just x free, then

$$\alpha(0) \wedge \forall x(\alpha(x) \rightarrow \alpha(x)) \rightarrow \forall x \alpha(x)$$

(PA4) $\forall x(0 \oplus x = x)$
(PA5) $\forall x \forall y(x \oplus \sigma y = \sigma(x \oplus y))$
(PA6) $\forall x(0 \otimes x = 0)$
(PA7) $\forall x \forall y(x \otimes \sigma y = \sigma(x \otimes y) \oplus x)$

The reader may easily note that (PA1)–(PA3) are possible formalizations of those axioms informally stated earlier. The additional postulates are necessary because we don't have in the theory the resources to prove that the functions representing the operations $\oplus$ and $\otimes$ can be defined (which requires the so-called Theorem of Recursion—see [End.77]). In short, with the help of this axiom, we can prove that the functions that define addition and multiplication exist and are unique. But, in first-order, logic this proof is not possible. According to Edmund Landau, this fact was observed for the first time by Grandjot [Lan.66, p.ix].

Let us give an example of a proof that will show the differences between the ways we usually proceed in abstract and in formal axiomatics. In group theory, we can prove the following theorem, known as the *cut rule*,[19] which can be stated in group theory as a theorem:

Theorem 2.3.1 (Cut Rule) *In every group $\mathcal{G} = \langle G, \circ \rangle$, for all elements $a, b, c \in G$, we have that $c \circ a = c \circ b$ entails $a = b$.*

Proof: (Informal) Let us suppose $c \circ a = c \circ b$. Since every element of G has an inverse still in G, then c has an inverse c'. So we can write $c' \circ (c \circ a) = c' \circ (c \circ b)$ once we can operate any two elements of G (in fact, $c', c \circ a,$

and $c \circ b$ are elements of G, so they can be operated—but note the lack of something here: How do we know that in writing c' before both terms in the equality that the equality still remains? This can be seen only with the resources of logic that follow). By the associative axiom, this last expression is equivalent to $(c' \circ c) \circ a = (c' \circ c) \circ b$. But, since c' is the inverse of c, it results from the last equality that $e \circ a = e \circ b$. So using the axiom of the neutral element, we get $a = b$. ∎

Now we present a detailed (and also boring) proof with all logical details. The main logical results to be used are the reflexive law of identity L6, namely, $\forall x(x = x)$, the substitutive law of identity L7 — that is, $x = y \rightarrow (\alpha(x) \rightarrow \alpha(y))$, where $\alpha(x)$ is a formula with a free occurrence of x and $\alpha(y)$ is obtained from $\alpha(x)$ by the substitution of y in some free occurrences of x and the inference rule Modus Ponens (MP). The proof goes as follows:

Theorem 2.3.2 (Cut Rule) *In every group G, for all elements a, b, $c \in G$, we have that $c \circ a = c \circ b$ entails $a = b$.*
Proof: (logic; in the right, we mention the used postulates)

1. $c \circ a = c \circ b$ (Assumption)
2. $(c \circ a = c \circ b) \rightarrow ((c' \circ (c \circ a) = c' \circ (c \circ a)) \rightarrow (c' \circ (c \circ a) = c' \circ (c \circ b)))$
   (1, L7), with $\alpha(a)$ being $(c' \circ (c \circ a) = c' \circ (c \circ a))$
3. $c' \circ (c \circ a) = c' \circ (c \circ a)$ (L6)
4. $(c \circ a = c \circ b) \rightarrow (c' \circ (c \circ a) = c' \circ (c \circ b)))$ (2, 3, propositional logic and MP, that is, the tautology $(\alpha \rightarrow (\alpha \rightarrow \beta) \rightarrow (\alpha \rightarrow \beta)))$
5. $c' \circ (c \circ a) = c' \circ (c \circ b)$ (1, 4, MP)
6. $(c' \circ c) \circ a = (c' \circ c) \circ b$ (5, G1)
7. $c' \circ c = e$ (G3)
8. $(c' \circ c = e) \rightarrow (((c' \circ c) \circ a = (c' \circ c) \circ b) \rightarrow (e \circ a = e \circ b))$ (7, L7)
9. $(((c' \circ c) \circ a = (c' \circ c) \circ b) \rightarrow (e \circ a = e \circ b)$ (7,8, MP)
10. $e \circ a = e \circ b$ (6,9, MP)
11. $e \circ a = a$ (G2)
12. $e \circ b = b$ (G2)
13. $(e \circ a = a) \rightarrow ((e \circ a = e \circ b) \rightarrow (a = \circ b))$ (11, L7)
14. $(e \circ a = e \circ b) \rightarrow (a = e \circ b)$ (12, 13, MP)
15. $a = e \circ b$ (10,14, MP)
16. $(e \circ b = b) \rightarrow ((a = e \circ b) \rightarrow (a = b))$ (15, L7)
17. $a = e \circ b \rightarrow a = b$ (12, 15, MP)
18. $a = b$ (15, 17, MP) ∎

The differences between informal (that is, not making the logical steps explicit) and formal proofs was discussed by Patrick Suppes in a very interesting and important chapter of his book [Sups.57, chap.7]. It is important to note that it would be practically impossible to present all mathematical deductions this way; for instance, in their masterpiece *Principia Mathematica*, Whitehead and Russell present the proof that $1 + 1 = 2$ on page 379! [WhiRus.10]. The amount of previously required definitions and proofs is overwhelmingly big. Anyway, it is beyond doubt that the logical analysis of mathematical proofs is important, so Proof Theory, the part of logic that deals precisely with this kind of analysis, is one of the fundamental subjects of modern logic. What is relevant here is that the aforementioned theorem does not hold only in a particular group, but it is valid for every group. This generalization is one of the main traits of AM, that is, the possibility of treating different fields of knowledge at once. What holds as a theorem for group theory holds for all groups, and the same can be said of all other abstract axiomatic systems.

Furthermore, without the development of modern logic and the full knowledge of the abstract and formal axiomatic systems, we could not arrive at strong results such as Gödel's, Tarski's, and others, like the independence of axioms in set theory which, as an example, gave us the possibility of finding alternative ('non-cantorian', in Cohen's words) mathematics. Thus the axiomatic method does involve heuristics. But this is a challenge for chapter 4. Before doing that, let us discuss a specific formal theory which shall have great relevance for our discussion on axiomatization and models: elementary set theory.

NOTES

1. Árpád Szabó attempts to understand why and how the transformation of mathematics into a discipline grounded in axioms and proofs takes place; see his mentioned works plus [Sza.78]. Patrick Suppes, on the other hand, in speaking about the origins of the axiomatic method, attributes it to Eudoxus in the fourth century BC [Sups.02, p.35], but we prefer to follow the (apparently) historically more detailed analysis developed by Szabó.
2. Today we no longer distinguish between axioms (anciently called common assumptions, valid for all sciences) and postulates (basic propositions of the specific science under investigation).
3. The proposition proposes to construct an equilateral triangle on a given segment. But the 'proof' uses a fact neither assumed nor proven before, namely, that two circles that appear in the proof intersect one another [Eucl.08].
4. A wonderful analysis of the origins of Greek mathematics is the book *The Beginnings of Greek Mathematics* by Árpád Szabó [Sza.78] as well as his mentioned papers.
5. According to Szabó, Aristotle did not believe that the fundamental principles of mathematics—that is, geometry—could be chosen arbitrarily, but that there would be "simple and natural starting point for mathematics" [Sza.78, p.227], [JonBet.10].
6. His axiomatics preceded Hilbert's by several months. The set of motions was characterized as a group of transformations acting on the set of points. See [Mar.93].
7. In fact, Newton's three laws can be taken as axioms of his system. But no explicit mention to axiomatization was made in his work.

8. There is a theory of *multisets* that enables 'sets' to have repeated elements [Bli.88]. There are also the *quasi-sets* whose elements may be indiscernible [FreKra.06].

9. For a good history of set theory, see [Dau.90].

10. This should be qualified, for it will be important for us latter. The domain (so as the universe and many other collections) are not *Zermelo-sets*, that is, sets in Zermelo's theory—supposed consistent—but nothing excludes that they can be sets of another set theory. The notion of set is not absolute, and depends on the theory we use.

11. For these 'known paradoxes', the reader may have a look at the introduction of [Men.97] and also in [FraBarLev.73].

12. The notion of structure shall be introduced precisely later. For now, let us take it intuitively as comprising a set or some sets and operations and relations over the elements of these sets, plus some distinguished elements—not all these objects may be present in every structure, of course.

13. Truly, there are formulations that do not use Hilbert spaces; see [Sty.02].

14. Here we are taking the words 'application', 'function', and 'mapping' as synonymous.

15. Generally speaking, a norm on a linear space is a mapping $\| \ \|$ from $\mathcal{V}$ in $\mathbb{C}$, which associates to any vector α a scalar $\|\alpha\|$ such that, for any vectors α and β and scalar x, we have (a) $\|\alpha\| \geq 0$ and $\|\alpha\| = 0$ iff $\alpha = \mathbb{O}$, (b) $\|x\alpha\| = |x|\|\alpha\|$, where $|x|$ is the modulus of x and (c) $\|\alpha + \beta\| \leq \|\alpha\| + \|\beta\|$. There are several different norms over a linear space, but we are interested in the one given by the aforementioned definition. A further remark: if x is a complex number $x = a + bi$, then $|x| = \sqrt{a^2 + b^2}$. The reader should notice that in the definitions given in this note, we are not making explicit differences between the operations among vectors and among scalars; by the way, this is the standard procedure we shall pursue from now on.

16. An alternative approach would be 'algebraic', seeing a logic as an ordered pair (an algebra) $\mathcal{L} = \langle F, \vdash \rangle$, where F is a set whose objects are called *formulas* and $\vdash$ is a function from $\mathcal{P}(F)$ to F, the deduction operation, satisfying well-known postulates.

17. Of course there are here also different ways of formulating $\mathcal{L}^1$, say by choosing a distinct set of logical symbols, but we shall omit the details.

18. This example is adapted from [Men.97, p.69], where further details and examples can be found.

19. Don't confuse with Gentzen's cut rule!

3 A Mathematical Background

We shall now outline a mathematical framework where the formal counter-part of scientific theories can be discussed—namely, the Zermelo–Fraenkel set theory, ZF, perhaps with the Axiom of Choice, ZFC. Since the relevance of this subject is in general not well known by the general philosopher of science, we will provide further explanations about the details of the theory in hopes that they will help the reader to agree with us on the importance of this kind of study to our subject.

We write ZF(C) to refer to both of the mentioned theories; when a distinction between these theories is required, we shall make explicit reference. Of course, other frameworks could be used instead; there are many options for expressing at least part of present-day mathematics, such as higher-order logics, as in Carnap [Car.58] (with some examples of empirical theories) or category theory [Ger.85].

For the sake of simplicity, and to follow common usage, we have chosen to treat ZF(C) as a first-order theory, and its language, presented in section 3.2, will be called $\mathcal{L}_\in$.[1] But before describing ZF(C), it will be profitable to discuss a little bit about the construction of axiomatic/formal systems.

3.1 THE PRINCIPLE OF CONSTRUCTIVITY

In his book *Ensaio sobre os Fundamentos da Lógica*,[2] da Costa calls the following methodological rule 'the Principle of Constructivity':

> The whole exercise of reason presupposes that it has a certain intuitive ability of constructive idealization, whose regularities are systematized by intuitionistic arithmetics, with the inclusion of its underlying logic.
> [Cos.80, p.57]

What does he mean by that? The explanations appeared earlier in his book, and we base our explanation on his ideas, with the risk of not being able to do justice to much of the rich and clear contents of the book. In this section, when we make reference to a certain page only (say, by writing 'page 51'), it refers to the mentioned book by da Costa. To speak a little bit of such matters is of course quite important, for we intend to present a formal system (ZF) using expressions such as 'infinite set of individual variables', for instance, which presuppose a previous notion of the concept of 'infinity'; hence the construction of set theory as a formal theory apparently presupposes at least part of the very mathematics it intends to ground.

But let us pay close attention to da Costa:

> Formal disciplines are essentially discursive. But the discourse develops itself in different levels, and each one of them must be understood, or intuited, as already noticed by Descartes. Even if one reasons symbolically and formally, the different elementary steps of the evolution of the discourse need to be clear and evident, for in the contrary there would be no reasoning, and one would not know what she is doing.
>
> (p.50)

In trying to systematize the empirical reality, we make use of our capacity of reasoning. We use intuition and other resources, such as previously learned theories, personal experiences, memory, imagination, expertise, and insights. In a certain sense, the present-day stage of the evolution of science is a result of not only our biological and cultural characteristics but also of the previous theories and scientific background. But even intuition is not enough. We need to systematize our intuitions, and in mathematics, the axiomatic method became the (apparently) best methodological tool, being extended to empirical science, as we have already seen. Well-developed sciences, or disciplines, even though they are not completely axiomatized or formalized yet, can be treated, from a mathematical perspective, in the same way as the formal sciences, being also essentially discursive. In other words, scientific knowledge, being essentially conceptual knowledge, needs discourse, and our 'discourse' needs language and symbolism (p.35) and, we can add, our inferences depend on logic.[3]

In a certain sense, we may say that our knowledge of a certain scientific domain is given by means of the elaboration of structures; we 'structure' the domain by gathering concepts, which may be relations and operations over a certain domain (or domains) of objects we are interested in. Even if we do it only informally, in reality, we are elaborating structures that concentrate what we think are the basic concepts we intend to use in our approach (we have already said something on this respect in the preface). Certainly, different scientists can choose different concepts. Then we can work 'informally' by providing the relevant definitions and theorems, or more formally by providing to these concepts suitable (even if informally stated) postulates we think may govern them with respect to our intended domain(s). Thus we may say that our scientific knowledge is conceptual *and structural*. Sometimes these 'postulates' are not even formulated explicitly. There are examples in mathematics (and much more in the empirical sciences) where a certain field is not presented axiomatically in the sense described earlier. For instance, consider analytic geometry as presented in a first undergraduate course. It has no axioms, properly speaking. The same happens to differential and integral calculus. These disciplines are developed from definitions to theorems. But, as Alonso Church has made clear, every axiomatic theory can be transformed into a 'definitional' theory, comprising just definitions and theorems. Thus, in a certain sense, both analytic geometry and differential calculus can be seen as 'axiomatic' too.

Going back to the point under discussion, we recall, as is well known, that in intuitionistic mathematics, we have an intuitive 'visualization' of the entities that interest us (p.50). This is essentially an *intellectual intuition*—an expression that intends to capture the idea that

> there cannot be immediate and evident knowledge without contemplation, without a look to the objects that interest us or, at least, of the conceptual relations which define them; in an analogous way, there is no intellectual contemplation which does not enable us to formulate direct judgments, linked to different levels of evidence.
>
> (p.51)

The reference to intuitionistic mathematics is grounded on the fact that, according to da Costa, it provides an intuitive 'visualization' of the entities that interest us (p.52), contrarily to standard mathematics, where such an intuition does not need to be available (p.54). He illustrates the issues with examples of the following kind: we have no clear 'vision' either of transfinite cardinals or of the totality of the real numbers, but only an intuition of the system of relations that implicitly define these concepts by means of axiomatic systems (p.54). This 'intuitive visualization' provides us with an *intuitive pragmatic nucleus*, and we may say that, based on this nucleus, we articulate a kind of algebra (da Costa doesn't use this word in this context) that enables us to compose them, operate with them, and so on, going to more sophisticated and sometimes not intuitively evident conceptualizations. This way, we go out of the intuitive nucleus, and in flying so high, the axiomatic method is our 'autopilot', which enables us to fly in domains where our intuitions do not help us much.[4]

In this sense, we begin by describing a formal system using this intuitive nucleus of finitist and constructive nature; as da Costa says, "it is today universally accepted that there cannot be formalized arithmetics without intuitive arithmetics" (p.57). It is this informal and intuitive handling of symbols and concepts that, at first glance, enable us to make reference to the tools we need to characterize our axiomatic/formalized theories. Thus it is in this sense that we formulate the language of ZF(C) described in the next section; in saying, for instance, that we have a denumerable infinite collection of individual variables, we could say that our language encompasses two symbols, x and $'$, and the variables would be expressions of the form $x, x', x'', \ldots$, and the same for the individual constants. This way, we avoid speaking of idealized concepts (in Hilbert's sense, such as 'denumerable many') and keep ourselves in a constructive setting. Interesting as it is, we had enough for this kind of discussion.

It is interesting that similar ideas are also advanced by Kenneth Kunen in his book *The Foundations of Mathematics* [Kun.09], which refer specifically to logic. Speaking about how set theory is developed, Kunen says that

> [y]ou don't need any knowledge about infinite sets; you could learn about these as the axioms are being developed; but you do need to have some basic understanding of finite combinatorics even to understand what statements are and are not axioms. [...] This basic *finitistic reasoning*, which we do not analyze formally, is called the *metatheory*. In this metatheory, we explain various notions such as what a formula is and which formulas are axioms of our *formal theory*, which here is ZFC.
>
> [Kun.09, pp.28–9]

We can interpret these words as indicating that we start with intuitive notions necessary, say, to distinguish between two different symbols (such as a and b). So we are presupposing things like the intuitive meaning of the number two as the idea of 'different' things. Then we start to combine these symbols in a way that enables us to elaborate sophisticated mathematical languages, such as that of ZF(C). Then, in a second stage, we work *inside* ZF(C) and, with its tools, we can *reconstruct* the steps we have gone through previously only at an informal level and, in particular, explain rigorously what *two* may be taken to be. In this sense, as says Kunen, "formal logic must be developed *twice*" (ibid., p.191).

The metatheory, of course, can also be treated formally, but in order to do that we would need to have available once again, as a prerequisite, an informal *metametatheory* and so on. But let us turn now to ZF and ZFC set theories.

3.2 THE ZF(C) SET THEORIES

Here we shall present two theories, ZF and ZFC. Some redundancies can be found by the reader with things already exposed before. But our aim is to keep this chapter independent, as it were, so we shall not bother with these repetitions. The reader should not be confused by the terminology: ZF is the first-order Zermelo–Fraenkel set theory *without* the Axiom of Choice, while ZFC is the theory obtained from ZF by adding to it the Axiom of Choice. The underlying formal language is the same for both theories, and they differ just by the mentioned axiom, so most of the developments may be carried on for both theories alike, with the relevant differences being mentioned when necessary. Important to note, as stated by Gödel and Cohen, the Axiom of Choice can be neither proved nor disproved from the axioms of ZF (supposed consistent).

The language $\mathcal{L}_\in$ of 'pure' ZF(C) (that is, not comprising *Urelemente*) has the following categories of primitive symbols:

    (i) The propositional connectives: $\neg$ and $\rightarrow$
    (ii) The universal quantifier: $\forall$
    (iii) Two binary predicates : $=$ and $\in$
    (iv) Auxiliary symbols: left and right parentheses: (and)
    (v) A denumerable infinite set of individual variables: $x_1, x_2, \ldots$, which in the metalanguage we will denote (provided that there is no risk of ambiguity) by x, y, z, w, u, v, etc.

The language $\mathcal{L}_\in$ is a denumerable language (its primitive symbols, given in the list presented earlier, can be enumerated) and its syntax, to be introduced now, is recursive, which intuitively means that we can program a computer to recognize whether a finite sequence of primitive symbols is a formula in the sense defined in the following section. First, let us say that the only terms of this language are the individual variables.

Definition 3.2.1 (Formulas) *The class of* formulas is defined recursively as follows:

    (i) If x and y are individual variables, then expressions of the form $x \in y$ and $x = y$ are atomic formulas.
    (ii) If α and β are formulas, so are all expressions of the form $\neg\alpha$, $(\alpha \rightarrow \beta)$ and $\forall x\alpha$, where x is an individual variable.
    (iii) There are no other formulas.

Frequently, we omit parentheses when the reading does not lead to ambiguity, and we reintroduce them in order to avoid ambiguity, according to standard conventions (see [Men.97]). We remark that α, β, and other symbols not listed earlier do not make up part of the basic language; they may be seen as *metavariables*. In this case, small Greek letters stand for formulas, and big Greek letters (as Γ, Δ) stand for collections of formulas. A further remark: In general, we do not work with $\mathcal{L}_\in$ properly speaking, but with definitional extensions of it, that is, with languages comprising more symbols than the primitive ones, for instance, those given by the next definition, so as with other common symbols, like $\subseteq$, $\aleph_0$, $\mathbb{R}$, $\emptyset$, and so on.[5] Anyway, we shall continue to speak about $\mathcal{L}_\in$ as the language posed earlier or one of its definitional extensions.

Definition 3.2.2 (Abbreviations) *Some abbreviations, given in the metalanguage, are, in order,*[6]

(i) $\alpha \wedge \beta := \neg(\alpha \rightarrow \neg\beta)$
(ii) $\alpha \vee \beta := \neg\alpha \rightarrow \beta$
(iii) $\alpha \leftrightarrow \beta := (\alpha \rightarrow \beta) \wedge (\beta \rightarrow \alpha)$
(iv) $\exists x\alpha := \neg\forall x\neg\alpha$
(v) $(\forall x \in y)\alpha := \forall x(x \in y \rightarrow \alpha)$
(vi) $(\exists x \in y)\alpha := \exists x(x \in y \wedge \alpha)$

In the expression $\forall x\alpha$, the formula α is called the *scope* of the quantifier (the same holds for $\exists$, since it is just an abbreviation). A variable x in a formula α in the scope of a quantifier $\forall x$ or $\exists x$ is said to be a bound variable. A *free occurrence* of a variable x in a formula α is an occurrence of x lies outside the scope of a quantifier $\forall x$ (idem for $\exists x$). In this case, x is said to be *free* in α. We frequently write $\alpha(x_1, \ldots, x_n)$ to indicate that all free variables of α occur in the list $x_1, \ldots, x_n$. A *sentence* is a formula without free variables.

Let $\alpha(x)$ be a formula of $\mathcal{L}_\in$ in which x is the only free variable. We shall call this formula a *condition*. In the intuitive theory of sets (which admits as 'sets' whatever collection of objects you wish), we have the following principle, called the *Axiom of Comprehension for Classes*; namely, given a condition $\alpha(x)$, there exists a collection (called a *class*) which consists exactly of those elements that satisfy the condition. We denote this class by

$$\{x : \alpha(x)\}, \tag{3.1}$$

which intuitively means 'the class of those x that obey the condition $\alpha(x)$'.

As is well known, it is this principle that originates the paradoxes in the intuitive set theory, such as Russell's. Just take as the condition the assertion that a set does not belong to itself (that is, $\alpha(x) \leftrightarrow x \notin x$); then, if we take the formed collection to be a set, there exists the set of all sets that do not belong to themselves, which belongs to itself if and only if it does not belong to itself.[7] So it is better to distinguish between sets and classes from now on; sets will always be relative to a set theory, so Russell's just mentioned class $\{x : x \notin x\}$ is not a set of ZF(C) (supposed consistent). The same happens with some 'large' classes, such as the class of all groups, of all vector spaces, of all models of a scientific theory, which are also not sets in ZF(C). These collections are very large to be sets of these theories (but they can 'exist', for instance, in some stronger set theories, as we shall see later). So what is a set? It depends on the axioms we use. In the next section, we shall see the postulates that determine which are the sets of ZF(C).[8] Thus, to summarize, we should be careful in saying ZF-sets, ZFC-sets, and so on. In principle, nothing prohibits Russell's class from being a set of some particular set theory; for instance, it is a PZF-set, where PZF stands for *paraconsistent Zermelo–Fraenkel set theory* (see da Costa et al. [CosKraBue.06]). To summarize, given a set theory T, we may have T-sets and other collections. The postulates of T provide the grounds for the characterization (implicit definition) of the sets *of* T.

3.2.1   The Postulates of ZF(C)

Let $\alpha(x)$ be a formula of $\mathcal{L}_\in$, where x is free. As we have seen, the collection of the objects satisfying $\alpha(x)$ is written $\{x : \alpha(x)\}$ and is called a *class*. The question is, which classes are sets? In other words, what conditions $\alpha(x)$ determine a set? The answer depends on the axioms we use, as we have already commented. In ZF(C) we shall have some collections that will deserve to be named *sets*, while others will not. Thus the postulates that follow determine which are the sets of ZF(C)[9]. There are different set theories (in fact, potentially infinitely many). One of the most well known is the system NBG, the von Neumann, Bernays, and Gödel set theory. In this theory, the basic objects are classes (see [Men.97, chap.4]). But some of them are sets, namely, those classes that belong to other classes. Those classes that do not belong to other classes were termed *proper classes* by Gödel. The sets of NBG and the sets of ZF(C) are essentially the same objects (NBG is a conservative extension of ZF(C)). As a main difference, however, notice that in ZF(C) we do not officially quantify over classes, while in NBG we do.

The logical postulates of ZF(C) are those of the first-order logic with identity, that is, being α, β, and γ formulas, the following schema provide the postulates:[10]

(1) $\alpha \to (\beta \to \alpha)$
(2) $(\alpha \to (\beta \to \gamma)) \to ((\alpha \to \beta) \to (\alpha \to \gamma))$
(3) $(\neg\alpha \to \neg\beta) \to ((\neg\alpha \to \beta) \to \alpha))$
(4) $\alpha, \alpha \to \beta/\beta$ (Modus Ponens)
(5) $\forall x\alpha(x) \to \alpha(t)$, where t is a term free for x in $\alpha(x)$
(6) $\forall x(\alpha \to \beta(x)) \to (\alpha \to \forall x\beta(x))$, if x does not appear free in α
(7) $\alpha/\forall x\alpha$ (Generalization)
(8) $\forall x(x = x)$
(9) $u = v \to (\alpha(u) \to \alpha(v))$, where u and v are distinct terms of $\mathcal{L}_\in$

These are the axioms of the underlying logic, which is the classical first-order predicate logic with identity. The specific postulates are the following:

(ZF1) [Extensionality] $\forall x\forall y(\forall z(z \in x \leftrightarrow z \in y) \to x = y)$
   This is just stating that sets with the same elements are identical, the same set. The converse of the postulate (obtained by reverting the arrow) is a consequence of the logic axioms, in special of (9).
(ZF2) [Pair] $\forall x\forall y\exists z(\forall w(w \in z \leftrightarrow w = x \lor w = y))$
   In the metalanguage, this set will be written $z = \{x, y\}$. Obviously, the introduction of such names (new symbols, as mentioned earlier) is allowed because we can, in each case, prove existence (by the pair axiom ZF2, in this case), and unicity (by the extensionality axiom ZF1). The same remarks hold for the new symbols defined in what

follows. A particular case of ZF2 is obtained when $x = y$. In this case, we write $z = \{x\}$ and call it *the unitary of x*, which is unique to ZF1.

(ZF3) [Separation Axioms] If $\alpha(y)$ is a formula of $\mathcal{L}_\in$ with y and $z_1, \ldots z_n$ as free variables in which x does not appear, then for each α, the following is an axiom: $\forall z_1 \ldots \forall z_n \forall w \exists x \forall y (y \in x \leftrightarrow y \in w \wedge \alpha(y))$. In particular, when α has only y as its free variable, then $\forall w \exists x \forall y (y \in x \leftrightarrow y \in w \wedge \alpha(y))$ is an axiom.

Taking $\alpha(x)$ as $x \neq x$, and applying the separation schema on a set w whatever, we get a set with no elements, which using the axiom ZF1 we can show is unique. This set is called an 'empty set' and denoted by $\emptyset$. The set x of postulate ZF3 is written (in the metalanguage) as $x = \{y \in w : \alpha(y)\}$. This poses a fundamental difference between this set and a class as given by (3.1). Here the elements that belong to x are 'separated' from an already given set w, while in (3.1) they are not coming from a previous set. Thus axiom (ZF3) is called the postulate of the limitation of size (of sets), and it is due to Zermelo.

(ZF4) [Union] $\forall x \exists w \forall z (z \in w \leftrightarrow \exists y (z \in y \wedge y \in x))$. The set w is written $\bigcup x$. In particular, we write $u \cup v$ for $\bigcup \{u, v\}$, that is, when x has just two elements, u and v.

Definition 3.2.3 (Subsets) *For any sets u and v, we pose $u \subseteq v := \forall w (w \in u \rightarrow w \in v)$ and say that u is a subset of v. Furthermore, we write $u \subset v$ to mean $u \subseteq v \wedge u \neq v$. In this case, we say that u is a proper subset of v.*

(ZF5) [Power Set] $\forall x \exists y \forall z (z \in y \leftrightarrow z \subseteq x)$. The set y is written $\mathcal{P}(x)$, the power set of x.

Let $\alpha(x, y)$ be a formula. We say that it is *x-functional* if for any x there exists just one y such that $\alpha(x, y)$ holds.[11] We say this by writing $\forall x \exists! y \alpha(x, y)$. Then we have

(ZF6) [Substitution Axioms]

$$\forall x \exists! y \alpha(x, y) \rightarrow \forall u \exists v \forall y (y \in v \leftrightarrow \exists x (x \in u \wedge \alpha(x, y))).$$

Intuitively speaking, the axiom (in reality, an *axiom schema*, for it gives us an axiom for each functional formula $\alpha(x, y)$ we consider) says that the image of a set by a function (as defined by a functional formula) is also a set.

Definition 3.2.4 (Sucessor) *The sucessor of a set x is termed x^+ and defined as follows:*

$$x^+ := x \cup \{x\}.$$

(ZF7) [Infinity]

$$\exists x(\emptyset \in x \land \forall y(y \in x \to y^+ \in x)).$$

By paying attention to the fact that $\emptyset^+ = \emptyset \cup \{\emptyset\} = \{\emptyset\}$, $\{\emptyset\}^+ = \{\emptyset, \{\emptyset\}\}$, etc., we realize that the set postulated to exist by the infinity axiom contains the empty set, its sucessor, and the sucessor of all its elements. Such a set is called *inductive*. The intersection of all inductive sets is defined to be the set of natural numbers.

Definition 3.2.5 (THE SET OF NATURAL NUMBERS) *The set of* natural numbers *is the least inductive set, denoted by ω.*

Such a set, the least inductive set, is formed by taking the intersection of all inductive sets; in informal mathematics, it is denoted by 'N'. Now, following von Neumann, we offer the next definition.

Definition 3.2.6 (THE NATURAL NUMBERS) *The (von Neumann) natural numbers are defined as follows:*

$0 := \emptyset$
$1 := 0^+ = \{\emptyset\} = \{0\}$
$2 := 1^+ = \{0, 1\}$
$3 := 2^+ = \{0, 1, 2\}$
. . .
$n := \{0, 1, \ldots, n - 1\}$
. . .

Thus, due to ZF7, we can write

$$\omega := \{0, 1, 2, 3, \ldots\}.$$

(ZF8) [Regularity or Foundation]

$$\forall x(x \neq \emptyset \to (\exists y \in x)(\forall z \in x)(z \notin y)).$$

This axiom says that any non-empty set x has an element whose intersection with x is the empty set. That is, such an element has as elements no element of x. Thus, in particular, x cannot have itself as an element. Let us prove this result with some care.

Theorem 3.2.1 *Due to the axiom of regularity, there exists no set having itself as an element.*
Proof: Let us suppose that $x \in x$. Then $x \neq \emptyset$. Since $x \in \{x\}$, we have $x \in x$ $\cap \{x\}$ (let us call this result $(\star)$). By (ZF8), there exists $y \in \{x\}$ such that $y \cap \{x\} = \emptyset$. But, since $\{x\}$ is unitary, the only element to be chosen is x itself.

Hence we must have $x \cap \{x\} = \emptyset$, which is contradictory to $(\star)$. So our hypothesis that $x \in x$ must be rejected.

It is important to say that the axiom of regularity is essential in this proof. If the theory does not assume it—it can be proven that ZF8 is independent of the remaining axioms of ZF(C)—[12] then this possibility remains open. In the same vein, without regularity, there may exist *extraordinary sets* (a term coined by Mirimanov in 1919), that is, sets involving chains of the following kind: $x \in x_n \in \ldots \in x_2 \in x_1 \in x$.

By adding the next axiom, we get the theory ZFC

(ZF9) [Axiom of Choice]

$$\forall x((\forall y \in x) \rightarrow y \neq \emptyset) \wedge (\forall t \in x)(\forall z \in x)(t \neq z$$
$$\rightarrow t \cap z = \emptyset) \rightarrow ((\exists t \subseteq x)(\forall u \in x) \rightarrow \exists v(t \cap u = \{v\}))).$$

The intuitive explanation of this axiom is given a non-empty set whose elements are also non-empty sets, and then there exists a set (the choice set) formed by just one element of each of the elements of the given set. There are many formulations equivalent to this one.

We could formulate these axioms without employing any of the defined symbols such as $\subseteq$, $\cup$, and others. But we have used them here to keep the text more easily readable. The Axiom of Choice is considered the second most famous axiom of all mathematics, losing only to Euclid's parallel postulate. There are excellent books on its history and importance (see for instance [Moo.82]; [FraBarLev.73, ch.2]).

3.2.2 A Matter of Terminology

We can consider several theories different from the aforementioned axiomatics. Let us fix some terminology:

1. ZF is the theory without the Axiom of Choice
2. ZFC is the theory encompassing all axioms noted earlier.
3. ZF$^-$ is ZF minus the axiom of regularity and the same for ZFC$^-$.
4. Z^p is the theory obtained by dropping the axioms of regularity and replacement. This is essentially the 'pure' Zermelo's set theory (without ur-elements or *Urelemente*).
5. ZF* is the theory obtained from the aforementioned axiomatics (without choice) by adding the Axiom of Inaccessible Cardinals to be explained in the next section; similarly, we get ZFC*.

We remark that all these theories have the same language and underlying logic. Their differences reside in the specific axioms.

3.2.3   Inaccessible Cardinals

There are some collections whose existence cannot be proved in ZF(C), supposing this theory is consistent.[13] Among them, we mention just the statement to the effect that there exists a strong inaccessible cardinal. We don't need to consider those statements here (but see the footnote at the end of this chapter); we mention them just to acknowledge their importance, for by adding the mentioned axiom, we get a theory ZF(C)* where we can prove the consistency of ZF(C).

A cardinal λ is strongly inaccessible if it satisfies the following conditions: (i) for every cardinal β, if $\beta < \lambda$, then $2^{\beta} < \lambda$; (ii) if $\{\beta_i\}_{i \in I}$ is a family of cardinals such that $\text{card}(I) < \lambda$ and $\beta_i < \lambda$ for every $i \in I$, then $\sup\{\beta_i : i \in I\} < \lambda$. It is enough to acknowledge that formulas expressing these facts are available and can be added to ZF(C) in order to get a stronger theory in which a model of ZF(C) can be built; we shall return to this problem soon. Other postulates could be introduced with the same finality, such as those postulating the existence of a *universe*, but we shall not consider them here (but see the next chapters and [BrigCos.71]).

3.2.4   Informal Semantics of $\mathcal{L}_{\in}$

From the formal point of view, the symbols of $\mathcal{L}_{\in}$ have no meaning. But, intentionally, we usually accept that they make reference to the objects of a certain intuitive non-empty domain of objects we conceive as representing *sets*, but in principle, they could be of any 'nature' we wish. Let us be a little bit more precise.

Definition 3.2.7 A structure *for interpreting* $\mathcal{L}_{\in}$ *is an ordered pair* $\mathfrak{A} = \langle D, \xi \rangle$, *where D is a non-empty set and ξ is a binary relation on D.*

Due to Gödel's second incompleteness theorem, the consistency of ZF(C) cannot be proved with its own resources. We need a stronger theory where we can build a model of ZF(C), such as ZF(C)* mentioned earlier. Thus the structure $\mathfrak{A}$ of the previous definition cannot be a set of ZF(C). The construction must be done, say, in ZF(C)* or in informal metamathematics (informal set theory).

The intended interpretation ascribes the *diagonal* of D—namely, the set $\Delta_D = \{\langle a, a \rangle : a \in D\}$—to the identity symbol of $\mathcal{L}_{\in}$ and the binary relation ξ to the membership relation. The expressions $\forall x \alpha(x)$ and $\exists x \alpha(x)$, where $\alpha(x)$ is a formula of $\mathcal{L}_{\in}$ in which x is the only free variable, mean 'for all elements of D' and 'there exists an element of D' such that $\alpha(x)$ holds, respectively.

One of the outstanding problems in the study of models of set theory is to find an interpretation that makes the postulates of ZF 'true' (in the intuitive sense). To begin with, let us reproduce here some examples of possible interpretations for $\mathcal{L}_{\in}$. Later we shall mention some more suitable ones. The first interpretation takes D as the set $\mathbb{Z}$ of the integers and ξ as the usual ordering relation $<$ on the integers. Thus the integers are now our 'sets', and $x \in y$ means $x < y$. Of course, this is an interpretation as we have described informally; to the predicate $=$ of identity we associate the set of all couples $\langle x, x \rangle$ with $x \in \mathbb{Z}$, and the connectives and quantifiers are interpreted as usual ($\exists x$ means 'there exists an integer x so that ...' and so on). It is easy to see that this structure models the axiom of extensionality; just substitute $<$ for $\in$ in the axiom to see that its translation to the language of the integers becomes a theorem of this theory. In the same way, we can show that the pair axiom also holds in this structure. But the theorem that asserts that there exists an empty set does not hold. In fact, let us consider the sentence (a formula without free variables)

$$\forall x_0 \exists x_1 (x_1 \in x_0). \tag{3.2}$$

The reader should recognize that (3.2) stands for a formula of $\mathcal{L}_\in$, which expresses the mentioned theorem. In the metalanguage, we could write $\forall x \exists y (y \in x)$. According to our interpretation, it means that *for any integer there exists an integer that is less than it*. It is easy to see that the translation of this sentence to the language of our interpretation (there exists an integer that is less than any integer) is false. Thus the given structure is not a model for ZF(C), for it does not model one of its postulates.

Now let us take another interpretation using $D = \mathbb{N}$ (the set of the intuitive natural numbers) and ξ being the order relation $<$ on such a domain. Now not only are the pair axiom and the extensionality axiom true according to this interpretation but also the theorem, asserting that there exists an empty set. But the new structure does not model all axioms of ZF(C), which can be checked by the reader. In fact, these interpretations are far from suitable for modeling all the axioms of ZF(C).

3.3 'MODELS' OF ZF(C), AGAIN

A model of a formal theory, as we already know, is an interpretation that makes its postulates true. Can we think of models of ZF(C) in this sense? As we have mentioned before, we can, but the corresponding structures will be 'sets' that are not ZF(C)-sets.

The consistency of ZF(C) can be proven only relative to another stronger theory whose consistency is then put also into question, and to answer whether this stronger theory is by its turn consistent, we will need a still stronger theory, and so on. An *absolute* proof of consistency requires that we prove that the theory does not entail two contradictory formulas. The two notions of consistency involved here—namely, relative and absolute—are also called syntactical (when absolute) and 'semantic' (when relative). More precisely, the syntactic (absolute) definition says that a theory T, whose language contains a negation symbol $\neg$, is consistent iff there is no formula α such that both α and $\neg\alpha$ are theorems of T. The semantic (relative) definition says that a theory is consistent if and only if it has a model (an interpretation that satisfies the axioms of T). Semantic consistency implies syntactic consistency and Henkin's Completeness Theorem (which holds in a restricted form also for higher-order languages) establishes that syntactic consistency is enough for a theory to have a model.

The reason we cannot prove the relative consistency of ZF(C) within the theory itself was mentioned earlier, and is due to Gödel's second incompleteness theorem. But, being consistent, ZF(C) has 'models'. The model that interests us here is called *the well-founded cumulative hierarchy of sets*, or *the von Neumann universe*, termed $\mathcal{V}$. The construction of $\mathcal{V}$ is made on transfinite recursion on the class *On* of the ordinals (which is also not a ZF(C)-set),

$$On = \{0, 1, 2, \ldots, \omega, \omega + 1, \ldots, \omega 2, \omega 2 + 1, \ldots, \omega^2, \ldots\}. \tag{3.3}$$

The construction goes by defining a hierarchy V_α ($\alpha \in On$) as follows:

$V_0 = \emptyset$
$V_1 = \mathcal{P}(V_0)$
$\vdots$
$V_{n+1} = \mathcal{P}(V_n)$
$V_\lambda = \bigcup_{\beta < \lambda} \mathcal{P}(V_\beta)$, for λ a limit ordinal[14]
$\vdots$
$\mathcal{V} = \bigcup_{\alpha \in On} V_\alpha$

In terms of structures, we may write $\mathcal{WF} = \langle \mathcal{V}, \in \rangle$, with 'WF' standing for 'well-founded'; that is, the sets in the universe obey the axiom of foundation.

Now it is a mathematical exercise to show that the axioms of ZF(C) are 'true' (in the intuitive sense) in this structure, which is represented in Figure 3.1.

The remarks that follow will make reference to both universes. First of all, it should be noticed that any V_α is a transitive class. Now let us consider the following question: given a certain ordinal α, which axioms of ZFC are satisfied in $\langle V_\alpha, \in \rangle$, where $\in$ the membership relation relativized to V_α (that is, $x \in y$ means $x \in y$ and $x, y \in V_\alpha$)? It is interesting that we can prove the following results:[15]

(i) Any V_α is a model of the following axioms: extensionality, separation, union, power set, choice, regularity.

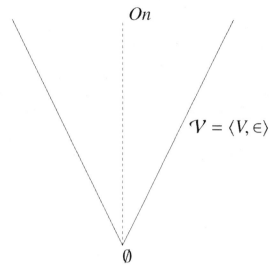

$$On$$

$$\mathcal{V} = \langle V, \in \rangle$$

$$\emptyset$$

Figure 3.1   The well-founded 'model' of *pure* ZF(C)

(ii) For the pair axiom, we need α as a limit ordinal. This can intuitively be understood for the set of two objects is of a higher level than the levels of the objects.

(iii) For the infinity axiom, we need that α be a limit ordinal greater than ω, for instance, $\omega 2$.

(iv) For the replacement axioms, we need more, for instance, an *inaccessible cardinal*.

(v) The whole $\mathcal{V}$ is a model of ZF(C).

Remember that we call Z^P the Zermelo set theory, which is ZF without the substitution (replacement) axioms. All the levels V_α can be constructed in Z^P as well. But, from the aforementioned results, we see that V_{ω^2} is a 'model' of Z^2, since ω^2 is a limit ordinal. Then, if consistent, Z^2 cannot admit V_{ω^2} as one of its sets, for in this case, we would be against Gödel's second incompleteness theorem. Reasoning in the same vein, we cannot prove the existence of inaccessible cardinals within ZF(C). Thus we can't prove the existence of $\mathcal{V}$, the whole universe of sets, within ZF(C). Of course, we could think that the universe can be constructed within a stronger theory, say ZF^*, which is ZF plus an axiom that says that there exists an inaccessible cardinal (according to the terminology introduced earlier). This is true, but we would just transfer the problem to this stronger theory, for we also don't know what *its* notion of set means, and a model for it would be needed. This problem raises interesting philosophical questions, due to the fact that, if consistent, a set theory such as ZF(C) (formulated as a first-order theory) is not *categorical*.[16]

We have made such a digression on set theory and on its 'models' just to show to the reader that there are differences in speaking of certain structures as models of certain theories and 'models' of the set theories themselves. The models of both mathematical and scientific theories, such as groups, vectors spaces, Euclidean geometry, metric spaces, differentiable manifolds, classical mechanics, and relativity theory can be constructed within Z^p, ZF, or ZFC, depending on the particular theory. But the 'models' of these theories themselves cannot be constructed within themselves (supposing them consistent). In the next section, we shall discuss a particular case involving quantum mechanics.

3.4   URELEMENTE

Abraham Fraenkel has shown that the ur-elements are not necessary for the foundations of mathematics, and his ideas (just as Skolem's) have led to a 'pure' set theory whose stages begin from the empty set seen earlier. But, for applications in the empirical theories, it seems that we need entities that are not sets, the ur-elements, or simply *atoms*. Here we are supposing, of course, as it seems reasonable, that physical objects such as chairs, molecules of a gas, elementary particles, and so on are not sets, so they can be reasonably represented by atoms.

A typical set theory with atoms is termed ZFU (Zermelo–Fraenkel with *Urelemente*), perhaps encompassing the Axiom of Choice. Here we will not write ZFC(U). The universe of ZFU (drawn in Figure 3.4 can also be constructed by transfinite recursion on the ordinals by assuming the existence of a non-empty set U of *Urelemente*, as follows:

$$V_0 = U$$
$$V_1 = V_0 \cup \mathcal{P}(V_0)$$
$$\vdots$$
$$V_{n+1} = V_n \cup \mathcal{P}(V_n)$$
$$V_\lambda = \bigcup_{\beta < \lambda} \mathcal{P}(V_\beta), \text{ for } \lambda \text{ a limit ordinal}$$
$$\vdots$$
$$\mathcal{V} = \bigcup_{\alpha \in On} V_\alpha$$

The axioms of ZFU are similar to those above, but must be modified accordingly in order to admit the existence of distinct *Urelemente*. Let us present these axioms rather briefly. The language of ZFU, which we call $\mathcal{L}_U$ has, in addition to the primitive vocabulary of $\mathcal{L}_\in$, an additional unary predicate symbol U so that $U(x)$ says that x is an *Urelement*. So we have the following definition:

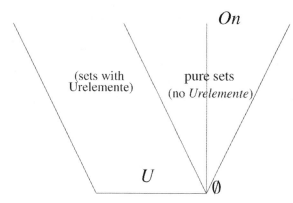

Figure 3.2 The universe of sets with *Urelemente*. *On* is the class of ordinals, and *U* is the set of atoms (*Urelemente*)

Definition 3.4.1 $\mathscr{C}(x) := \neg U(x)$

Intuitively speaking, $\mathscr{C}(x)$ says that x is a set. We use restricted quantifiers in the following sense: $\forall_\mathscr{C} x\alpha(x)$ means $\forall x(\mathscr{C}(x) \to \alpha(x))$, and $\exists_\mathscr{C} x\alpha(x)$ means $\exists x(\mathscr{C}(x) \land \alpha(x))$. The same can be done for other defined predicates (see axiom ZFU4). The logical postulates are those of the first-order classical logic with identity, and the specific postulates may be stated as follows:

(ZFU1)[Extensionality] $\forall_\mathscr{C} x\forall_\mathscr{C} y(\forall z(z \in x \leftrightarrow z \in y) \to x = y)$

Two sets are equal if they have the same elements (either sets or *Urelemente*).

In set theory, *Urelemente* is synonymous with *atoms* (objects that are not sets but which may be elements of sets). Thus sometimes the words 'atom' and 'atoms' are used there in this sense. This axiom enables the theory to be compatible with the existence of different atoms, other objects than the empty set (which is a set). Next, we shall speak a little bit about atoms.

(ZFU2)[Empty set] $\exists_\mathscr{C} x\forall y(y \notin x)$

The existence of an empty set termed $\emptyset$ and proved that it is unique by extensionality.

(ZFU3)[Pair] $\forall x\forall y\exists_\mathscr{C} z\forall w(w \in z \leftrightarrow w = x \lor w = y)$

As is usual, this set is written $z = \{x, y\}$.

Definition 3.4.2 $E(x) := \mathscr{C}(x) \land \forall y(y \in x \to \mathscr{C}(y))$

Which says that x is a set whose elements are also sets.

(ZFU4)[Union] $\forall_E x \exists_\mathscr{C} y \forall z (z \in y \leftrightarrow \exists w (w \in z \wedge z \in x))$

In the next two axioms, the definitions are as in the case of ZF(C).

(ZFU5)[Power Set] $\forall_\mathscr{C} x \exists_\mathscr{C} y \forall z (z \in y \leftrightarrow \mathscr{C}(z) \wedge z \subseteq x)$

(ZFU6)[Infinity] $\exists_\mathscr{C} x (\emptyset \in x \wedge \forall_\mathscr{C} y (y \in x \rightarrow y^+ \in x))$

(ZFC7)[Foundation/Regularity] $\forall_\mathscr{C} x (E(x) \rightarrow \exists y (y \in x \wedge y \cap x = \emptyset)$

Given that the elements of a set x are also sets, then x has an element that has no common element with x.

(ZFU8)[Replacement] Let $\alpha(x, y)$ be a formula where x and y are free and z is not free. Then

$$\forall x \exists! y \alpha(x, y) \rightarrow \forall x \exists_\mathscr{C} y \forall z (z \in y \leftrightarrow \exists w (w \in x \wedge \alpha(w, z))).$$

That is, if $\alpha(x, y)$ ascribe just one y for each x, then there exists the so-called α-image of any element of the domain.

(ZFU9)[Choice]

$$\forall_\mathscr{C} x (E(x) \wedge (\forall_\mathscr{C} y \in x)(y \notin \emptyset) \wedge (\forall z \in x)(\forall w \in x)(z \cap w = \emptyset) \rightarrow$$
$$\exists_\mathscr{C} t (\forall z \in t)(\exists s (z \cap t = \{s\})))$$

3.4.1   Some Comments

We could add to this axiomatic some further axioms, such as one of the following:

(ZFUa)[Non existence of atoms] $\forall x \mathscr{C}(x)$

(ZFCb)[Atoms are void] $\forall x \forall y (y \in x \rightarrow \mathscr{C}(x))$

The first one makes the theory essentially ZFC. The second one (which is implied by the first) says that atoms have no elements. This is interesting, for since we think of atoms as, say, physical objects, perhaps it would be better to think of them as composed of *parts* in a mereological sense.[17] Thus this axiomatics should be increased by a relation symbol for "part-whole" and the corresponding axioms for a mereology. But this is not our topic here.

An important result is that ZFU is equiconsistent with ZF(C); that is, one of them is consistent iff the another one is consistent too.

3.5   MORE ON THE 'MODELS' OF ZF(C)

We have remarked that ZF (Z^p, ZFC, ZFU), formulated as as first-order theory, is not categorical. Let us comment a little bit on this claim in order to see its consequences, even to physical theories.

We shall mention just a case, among several possible others, which could be obtained by using techniques such as Cohen's forcing, but which are out of the scope of this text. The importance of the 'models' of set theory for the philosophy of science may be explained as follows. A formalized theory (even of the empirical sciences), say by a Suppes predicate, is generally elaborated to cope with a certain informal given theory, as we have seen before. But in order to achieve this aim, we need to provide an *interpretation* to our formalism, generally by associating to it a structure built in set theory or, what is the same, in a 'model' of the set theory we use as our metatheory. But a set theory such as ZF (ZF(C), ZFU) has different models. If axiomatized as a first-order theory, it has even a denumerable 'model' due to the Löwenheim-Skolem theorem, and since, in general, we never know what model we are working in, we really will have difficulties granting that the interpretation we provide fits the intended concepts. We remark once more that despite the fact that we can think in sets and related concepts as if we are inside the 'standard model' of ZF(C), we cannot grant that this is really so, for we have no grounds for proving that such a 'model' exists. Let us consider a particular example concerning denumerable 'models' only.

Real numbers can be used in physics for representing time, probability measures, eigenvalues of self-adjoint operators, to parametrize certain functions, and so on. The physicist with no logical background has an intuitive idea of what real numbers are as well as what *sets* are. If pressed, a physicist may even make reference to some construction of the set of real numbers from the rationals, say by convergent sequences (Cauchy sequences) or by Dedekind cuts (see [End.77, pp.111ff], [Fra.82, pp.87ff]). Say these constructions are done within ZF(C). But suppose ZF(C) is consistent. Then it has 'models'. The first we can imagine may be called the *standard model*, which is more or less identified with a 'universe of sets' such as the universe $\mathcal{V}$ seen before. But, as we have seen, we cannot prove in ZF (C) that such a 'model' exists, so we don't have any evidence that our intuitive account of sets, that is, our intuitive idea of a set, is captured by the elements of such a universe. But there is more. As a first-order theory, ZF (C) is subjected to the Löwenheim-Skolem theorem, which implies two things: first, since any model of ZF(C) must satisfy the axiom of infinite, ZF(C) will have an infinite 'model' (recall that 'model' is being written in quotation marks to emphasize that it is not a set of ZF(C) but a proper class). Then, by the *upward* Löwenheim-Skolem theorem, ZF(C) has non-isomorphic models of any infinite cardinality. Second, let us fix in the other result entailed by this theorem: by the *downward* Löwenheim-Skolem theorem, having models, ZF(C) has a denumerable model.[18] What does this mean for empirical science?

Let us suppose that our theory T demands the real number system, as most physical theories do. Then it seems 'natural' to suppose that the set $\mathbb{R}$ of the real numbers is not denumerable, as Cantor showed in the nineteenth century with his famous diagonal argument. But in such a denumerable 'model', the set that (in the 'model') corresponds to the set of real numbers must be denumerable. This is of course puzzling, but there is not a contradiction here. As remarked by Skolem, this result, known as the 'Skolem paradox' is not a formal 'paradox', but just a result that goes against our intuition, showing that *within* ZF(C) (that is, a ZF(C)-set) there is no bijection between $\mathbb{R}$ and ω, but this does not entail that such a bijection does not exist *outside* ZF(C)—that is, as something that is not a ZF(C)-set. Thus if a physical theory demands a mathematical concept that, if standardly defined, presupposes the set of real numbers, how can we grant that it is non-denumerable? In other words, how can we ensure that, working within ZF(C) we are making reference to some 'model' that fits our intuitive claims? Unfortunately, this is not possible, and the best we can do is to fix a particular 'model' of ZF(C) when we need to give an interpretation of concepts, but we will always be subject to questionings. Indeed, as remarked, we can't show that even the so-called standard (intended) model of ZF(C) does exist!

On what concerns scientific theories, as remarked by da Costa, "we should never forget that set theories, supposed consistent, have non-standard 'models', thus any theory founded on them will have non-standard models too." [19] This entails that if we try to provide an understanding for the concepts we use (that is, by giving them an interpretation), we need to consider 'models' of ZF(C) and this poses a challenge: we never know in which model we are working so we really never know what our concepts actually mean. All we can do is *suppose* we are working in the standard 'model', where (apparently) sets are just as we imagine, that the real numbers cannot be enumerated, that there are basis for our vector spaces, and so on. The investigation of non-standard models for physical theories is still a novel domain of research, and it most likely will be a rich one.

NOTES

1. In fact, we could have used a different language, for instance, by adopting distinct primitive symbols and perhaps other postulates (which would be in most cases equivalent to the ones presented here). There is also the more relevant possibility of constructing a theory grounded on second-order logic, as Zermelo seems to have preferred—see [Moo.82, pp.267ff] —(or still higher-order) logic, and in this case, in fact, we should have different theories, for their properties would certainly be distinct.
2. There is a French version of this book, translated by Jean-Yves Béziau, called *Logique Classique et Non-Classique*, Paris, Flammarion, 1996.
3. In a broad sense, we use not only deductive inferences when elaborating a theory but also (at least) inductive and abductive, to use Peirce's teminology [Dou.11].
4. For instance, usually we reason more or less well in the domains where we suppose classical logic works, but if we need to go to other domains, logic needs to be made explicit.

5. We could add these new symbols for relations, operations, or individual constants to the language $\mathcal{L}_\in$, but with some care. The details about how this can be done is presented in [Sups.57, Chap.7].

6. The symbol ":=" means 'equal by definition'.

7. For more details, see the introduction of [Men.97].

8. Please be careful here: a group, a scientific theory such as classical particle mechanics, is described mathematically by structures that *are* sets. But the collections of *all groups*, of *all particle mechanics*, are not sets of ZF(C) supposed consistent.

9. However, notice that the axioms themselves do not contain the word 'set'. So, in this sense, the axioms only put some constraints in possible classes of objects satisfying them, with no warrant that every one of such classes will be close enough to our 'intended' model.

10. The meaning of the word 'schema' here is the following one: for instance, in the first one, we get a postulate (synonymous of 'axiom') by exchanging α and β by formulas of $\mathcal{L}_\in$, and in every case, the same Greek letters must be exchanged by the same formulas.

11. Intuitively speaking, $\alpha(x, y)$ is a formula that defines a function or mapping in the variable x.

12. A sentence S is independent of a theory T if the postulates of T prove neither S nor the negation of S. Remember that this is what happens with the famous parallel axiom in Euclidean geometry. The same will happen with the Axiom of Choice in ZF, as we have mentioned already.

13. Sometimes logicians bore the reader with details. But it is necessary to take into account that when we say that something (some set) 'exists' in ZF(C), we are assuming that this theory is consistent. If it is not, since its logic is classical logic, we can prove that any set (collection, class) belongs to it, which is something of course we would not like to admit. Hence the remark.

14. An ordinal α is a limit ordinal if there is no ordinal β such that $\alpha = \beta + 1$. Typical limit ordinals are ω, $\omega 2$, ω^2, etc.

15. Let M be a class (it may be a set) and F a formula of $\mathcal{L}_\in$. The formula F^M is called the *relativization* of F to M if it is obtained by substituting $\exists x \in M$ for $\exists x$, and $\forall x \in M$ for $\forall x$. Thus it says exactly what F says but concerning elements of M only. The results that follow are proven in [Fra.82, pp.298ff]; [End.77, pp.249ff]; [Fra.82, p.289].

16. A theory is categorical if all of its models are isomorphic.

17. Mereology, or the logic of parts and wholes, was developed initially by Stanislaw Lesniewicz, and has been studied from different perspectives ever since; see Simons [Sim.87].

18. Roughly speaking, the *downward* version of the theorem says that if a consistent theory T in a countable language has a model, then it has a denumerable model. The *upward* theorem says that such a theory, having infinite models, has models of any infinite cardinality.

19. Personal communication.

4 Criticism of the Axiomatic Method and Its Defense

There seems to be no doubt that the axiomatic method (AM) has achieved tremendous success in mathematics. Nowadays it would be difficult to find a field of pure mathematics not presented by employing this method (generally, as an abstract axiomatics). Real numbers, natural numbers, topological spaces, vector (linear) spaces, groups, rings, fields, etc., are presented axiomatically.[1] For instance, many mathematicians think that for some finalities it is easier to present, say, the real numbers as a complete ordered field than to construct the real numbers system as Dedekind cuts or as equivalence classes of Cauchy sequences (among other possibilities). Of course this depends on the aims and purposes of the mathematician. But it seems clear that the axiomatization of a certain field enables us to determine from a certain point of view those concepts and assumptions that could be taken as basic in the field, acknowledging also the resources available to the deductive apparatus employed (we mean the underlying logic—mathematicians usually assume something along the lines of classical logic in their deductions, so the use of classical reduction to the absurd, for instance, is available).

When a field of mathematics is presented axiomatically, it is in general as an abstract axiomatics; the field of real numbers, for instance, when presented as a complete ordered field, is nothing more than a particular structure of a certain general kind (the kind of complete ordered fields), or *species*, to use Bourbaki's terminology [Bou.68]. But we can easily realize that no standard mathematical book of analysis, algebra, etc., speaks of the underlying logic.[2] But despite the achieved success, the method was not taken for granted by all mathematicians and philosophers, not to speak of physicists. Several criticisms were advanced against axiomatization. In this chapter, we shall see some of them, and in the final part, we present a defense of the method.

4.1 DOES THE AM ENCOMPASS HEURISTICS?

One of the most interesting criticisms to the AM was advanced by Imre Lakatos in his *Proofs and Refutations* [Lak.76]. It should be remembered that in this book, as well as in other places, Lakatos tries to bring to mathematics the Popperian scheme of knowledge growing—namely, by 'proofs and refutations'. Lakatos holds that *informal (inhaltliche)* mathematics proceeds exactly as Popper proposed regarding the empirical disciplines. According to Popper, the scientist starts with a problem (P_1), provides a *tentative theory* (*TT*), which attempts to give us a tentative solution. Then tests are made and if something goes wrong, he or she tries to eliminate the errors (*EE*), which leads him or her to another problem, P_2.[3] Then the scheme is back to its starting point. The same happens in informal mathematics, says Lakatos. The mathematician starts with a problem, say Euler's conjecture that the relation between the number of vertices V, the number of edges E, and the number of faces F of polyhedra, in particular the regular ones, is given by Euler's formula $V - E + F = 2$. There is also a 'proof' of this conjecture. The 'proof' then faces counterexamples, showing that the formula does not apply to *any* polyhedra, say for that one formed by a cube with a hole whose intersection with the two opposite faces is a square (in this case, it was later noticed that $V - E + F = 0$).[4] The scheme goes as follows: given some evidence, the scientist makes a conjecture by means of a hypothesis (inside a certain scientific theory). The cases that agree with the conjecture simply *corroborate* it. But the conjecture may admit counter examples. In this case, some change is required in the conjecture (and in the theory). To Lakatos, the same happens in informal mathematics, as his example shows us. So, he says, (informal) mathematics proceeds by 'proofs' and refutations.

Axiomatic theories, on the other hand, says Lakatos, do not allow such revisions. What we derive from the axioms are the theorems of the theory, and they do not admit counterexamples (in the theory, supposed consistent). According to Lakatos, axiomatized mathematical theories, which he identifies with the formalist school headed by Hilbert (which as we see is not completely right, for we could make use of either concrete or abstract axiomatics without recurring to strict formalization), are tautological, just providing results (the theorems) which in a certain sense are already implicit in the axioms. A proof, in axiomatized mathematics, is nothing more than an unbreakable chain of mechanical procedures that go from premises to conclusions. Informal mathematics, on the contrary, as in Popper's account to natural sciences, proceeds by presenting criticisms to the conjectures, explanations, and discussions that make the conjecture more plausible and convincing due to the existence of counterexamples. So axiomatization does not involve heuristics, hypotheses, and explanations other than those already implicit in the axioms.

4.1.1 The Axiomatic Method and Heuristics

There is a reasonable sense according to which we can say that the AM involves heuristics. Following Suppes, we can distinguish between two kinds of axiomatics: *heuristic* and *unheuristic*. By heuristic axioms, says Suppes,

> I mean that the analysis yields axioms that seem intuitively to organize and facilitate our thinking about the subject, and in particular to formulate, in an ordinarily self-contained way, problems concerned with the phenomena governed by the theory and their solutions.
>
> [Sups.83]

That is, the heuristic axiomatics provides us a clear understanding of the field, this being understood as linked to our intuitions concerning the field under analysis. Peano's original axiomatics and Zermelo's set theory are typical examples, for they are in agreement with our intuitions. But Suppes also speaks of *unheuristic* axiomatics. He does not say that these types of axiomatization are devoid of value, but simply they are artificial, not intuitive, not representing "the kind of transparent and conceptually satisfactory solution we should aim at whenever possible" (ibid.). Suppes doesn't mention either Peano or Zermelo, but the axioms of the field of real numbers. The *construction* of the real numbers by Dedekind cuts or by equivalence classes of Cauchy sequences, he says, are unnatural to deal with, contrarily to the axioms of a complete ordered field, which he regards as "very natural and intuitive". The same happens with Kolmogorov's axioms for probabilities. The unheuristic example is Mackey's axioms for non-relativistic quantum mechanics. In fact, despite Mackey's formulation being very useful, there is no discussion about the motivations and it is quite artificial, for "the axioms taken literally present a wrong picture of how to think about physical problems in quantum mechanics".

Suppes links heuristics with intuitiveness, a vague concept. We think we have other arguments for defending the heuristic role of the axiomatic method by pointing out that only by axiomatizing arithmetics within a first-order logic were we able to discover the existence of non-standard models. The same can be said about first-order real number theory. Thus we have discovered things that were inaccessible to us without the strict application of AM; that is, without axiomatizing arithmetics, we would never be aware of non-standard natural numbers. The same can be said of many other fields; without axiomatics, we probably would never have arrived at non-Euclidean geometries, such as the study of distinct (non-equivalent) formulations of set theory, versions of set theory where the Axiom of Choice does not hold non-commutative and non-associative algebras, and so on. Even logic was made clear only after its formulation in axiomatic terms, starting with Aristotle. Thus we regard AM as heuristic it may lead to the discovery of many new fields of inquiry, although sometimes it may be quite artificial. In this sense, heuristic power needs not always coincide with intuitiveness.

4.2 TRUESDELL AGAINST THE 'SUPPESIANS'

As we shall see in more detail in the next chapters, Patrick Suppes and collaborators have developed a way of axiomatizing a scientific theory by assuming set theory (some set theory, but since it is not made explicit, we should take it in terms of his heuristics, that is, the most intuitive one—he speaks of *informal* set theory).[5] Set theory is a general framework where all usual concepts needed in mathematics (and in the usual physical theories) can be developed: functions, vectors, differential equations, manifolds, geometries.[6]

Recall that it was traditionally thought that the Received View of scientific theories required that every scientific theory should be axiomatized in a first-order language. Suppes [Sups.67], in particular, also advanced such a reading (which, as we have seen, is misguided; again see our discussion in chapter 1). As an alternative to such a restrictive framework, Suppes thought that logic and the basic mathematics should be taken for granted if we assume a set theory right from the start, as we shall see in chapters to come. Suppes didn't mention *which* set theory he was thinking about, but this is just what he intended to presuppose: any set theory able to express the necessary mathematical tools demanded by the theory being dealt with will do. Without the need of specifying a particular set theory, naïve set theory can be taken for granted. Hence by presupposing logic and the relevant mathematics, he presents a case of what we have termed 'abstract axiomatization'.

As we have already discussed briefly in chapter 1 and we will present with details in chapters 5 and 6, Suppes's technique allows us to represent a theory as a given class of models of a certain set-theoretical predicate (here, recall, models are understood in terms of mathematical structures). One of the most celebrated examples is classical particle mechanics (CPM), which McKinsey, Suppes, and Sugar presented in an axiomatized form in 1953 (see [Sups.02] for an updated version and also the next chapters).

CPM deals with 'particles' and the forces acting on them. But what are particles? In the spirit of abstract axiomatics, we should not pose this question. It is a primitive concept. CPM serves to deal with several distinct *models* of particle systems, such as the solar system, the molecules of a certain gas, and, in general, with whatever collection of objects that fulfill the conditions of the theory and make its postulates true. In each of these models, the objects of the domain are the particles. But CPM does not deal with interaction among particles, with collisions and other phenomena typical of what in physics is called *continuum mechanics*. And here lies the core of one of the most famous attacks on the axiomatic method as employed by Suppes: Truesdell's criticism.

Clifford Truesdell was an influent American mathematician who contributed to the foundations of several areas in the physical sciences, in particular in continuum rational mechanics. He was also the editor of the *Journal of Rational Mechanics and Analysis*, to which McKinsey, Sugar,

and Suppes submitted their papers on the axiomatization of classical particle mechanics [McSS.53], so Truesdell's opinion on the issue should be of
greatest relevance. One of the papers was published with a note by Truesdell, stating that

> [t]he communicator is in complete disagreement with the view of clas-
> sical mechanics expressed in this article. He agrees, however, that strict
> axiomatization of general mechanics—not merely the degenerate and
> conceptually insignificant special case of particle mechanics—is ur-
> gently required. While he does not believe the present work achieves
> any progress whatever toward the precision of the concept of force,
> which always has been and remains still the central conceptual
> problem, and indeed the only one not essentially trivial, in the founda-
> tions of classical mechanics, he hopes that publication of this paper
> may arouse the interest of students of mechanics and logic alike, thus
> perhaps leading eventually to a proper solution of this outstanding
> but neglected problem.

Later, in chap.39 of his book *An Idiot's Fugitive Essay on Science:
Methods, Criticism, Training, Circumstances*, titled 'Suppesian Stews'
[True.84], he considered once again Suppes et al.'s methods, providing a
long criticism, saying in particular that

> McKinsey, Sugar, & Suppes gave no example of how to use their par-
> adigm [that of presenting a set theoretical predicate, as it became clear
> in Suppes' works, as we shall see later] for any purpose in the practice
> or development of mechanics, and so far as I can learn, no-one has done
> so. While the Suppesians content themselves with admiring it for its
> perfection, specialists in mechanics have given in no heed, perhaps
> because it leaves out of account too much of the meat of the the me-
> chanics of Newton, Lagrange, and Cauchy—or, for that matter, of
> the content of any current textbook of mechanics for engineers.
>
> [True.84, p.539]

It is clear that Truesdell is criticizing the paper by McKinsey et al. for not
covering what he (correctly) considers the main part of mechanics—namely,
continuum mechanics—which deals with collisions and other concepts not
covered by particle mechanics. Part of his criticism is taken from a previous
book of Wolfgang Stegmüller [Steg.79], who has distinguished two versions
of the 'structuralist view' of theories in the version defended by W. Balzer
and others, which he calls 'the *Carnap approach*', and the version by
Suppes et al. termed the *Suppes approach*. Of course there are strong differences among these views that do not concern us here. What is relevant is that
Truesdell is complaining about the lack of adequate content of the proposal
of McKinsey et al. on what concerns the 'real mechanics'.

The question behind this criticism can be posed as follows: should an axiomatic version of a theory necessarily cover *every* aspect of the theory? The answer will lead us to a future section termed 'Does axiomatics really axiomatize?', so we will not discuss it in full here. We only advance our opinion that no, an axiomatic theory does not necessarily need to cover all the theory's subject area due fundamentally to two reasons: i) the scope of an informal theory (which supposedly is being axiomatized) is not well defined, and ii) this may not be of immediate interest to those who propose the axiomatics. The first item is obvious. The second can be enlightened by McKinsey's response to Truesdell, who had sent him a letter by Georg Hamel evaluating the papers by the three authors given on the following grounds:

> Thank you [Truesdell] so much for your letter of October 21st, with the enclosed report of Professor Hamel on the two papers by Sugar, Suppes, and me.[7] I should like to tell you that we have not been very impressed by Hamel's criticisms, and that we still want to publish these papers in their present form.
>
> In the first place, with regard to Hamel's criticism that our treatment is restricted to 'the most meager mechanics of points', it does not appear reasonable to us to object to a scientific paper on the ground that it has not accomplished something which the authors were not trying to accomplish: one does not criticize a paper on linear differential equations for not also covering non linear differential equations. We are of the opinion, moreover, that as a preliminary to any adequate treatment of the mechanics of extended bodies,[8] it is desirable (or perhaps even necessary) to present classical particle mechanics in a clear and precise form. In addition, such a presentation would be useful for an analysis of relativistic particle mechanics—and of continuum particle mechanics, both in classical and relativistic form.
>
> (quoted from [True.84, p.525])

They are correct. If that were not the case, no axiomatic presentation of classical propositional logic would be possible, on the grounds that it does not cover classical quantified logic, for instance. Furthermore, an axiomatic treatment of continuum mechanics, a topic addressed by Truesdell as lacking in the approach by the three authors, was developed by Walter Noll, a former student of Truesdell. It is interesting enough that Noll presented *three* versions of such a theory, and the reasons would be questionable by Truesdell's criterion: should the first two not be enough to cover all continuum mechanics? In this case, why Truesdell did not address similar restrictions to Noll's work? Well, this is a question to historians, and we wish just to point that these kind of criticisms to the axiomatic method are certainly misguided: Truesdell attacks the *method* by complaining about restrictions on the scope of *content* of the mechanics being axiomatized.

4.3   ARNOL'D AND THE BOURBAKI PROGRAM

The great Russian mathematician Vladimir Arnol'd famously made clear his opinion about the axiomatic method in general and about Bourbaki's program of seeing mathematics as the science of structure, grounded in that method [Arn.97]. Arnol'd was sanguine in criticizing AM and Bourbaki, calling him "the devil" of a tendency, which had flourished in the beginnings of the twentieth century (mainly due to Hilbert, he recalls for us), of considering that AM should be propagated in mathematics.[9] Arnol'd considered this "a self-destructive democratic principle", for it led mathematics to depart from physics and the other sciences. The strict application of this method, and surely he is referring here to what we have called *abstract axiomatics* in the previous chapter, is that it turns any calculus into a *blind calculation*, for the students are invited to acknowledge that the multiplication of standard numbers is commutative even if they know nothing about numbers.

Arnol'd suggests that blind calculations should not substitute content and that the intuition of the mathematician is something quite important to be dismissed in favor of purely logical deductions. In this sense, he is at least partially in agreement with Imre Lakatos and his criticism of formalization, claiming for intuitive mathematics. In our opinion, his understanding of the aims of the use of the axiomatic method are completely misguided. We believe that no one, Hilbert and Bourbaki included, aimed at substituting mathematics by blind calculations. Hilbert was right in proposing that the use of formal methods has a finality: mainly to avoid that 'intuitive' elements not part of the theory play some decisive role in our deductions. When such formalization efforts were launched, it was still very fresh in everyone's mind how geometric proofs were performed with the aid of geometrical figures, introducing unallowed elements into the proof (a typical well-known example is Euclid's Proposition 1).[10] The advantage of using the AM is precisely this: to provide the mathematician with solid grounds in the proofs. He or she knows what is being accepted, the methods of deduction that are allowable by the underlying logic, and then the kind of theorems he or she could obtain. Notice it was precisely this kind of recognition of the importance of making explicit our underlying assumptions that made possible intuitionistic mathematics, which does not accept indirect proofs, or paraconsistent mathematics, where some kind of 'contradictions' are to be accepted.

4.4   WORTH AXIOMATIZE?

The advantage of the AM, as posed in the last paragraph, seems to be clear. Once fully axiomatized, a theory becomes (at least in principle) a clear and objective object of study. We can know which are its basic (primitive) concepts, which are its fundamental laws (postulates), and if complete formalization is also considered, we can also know about its underlying logic. The logical structure of the theory becomes completely clear, for good or for bad. But perhaps one of the greatest possibilities opened up by formalization and full employment of the axiomatic method was that of meta-mathematical studies. During the twentieth century, as is well known, due to the process of formalization (say of arithmetics, set theory, and many other mathematical theories), we have gained a fine knowledge of the theories and their limitations. Without such a process, there would be no Gödel's incompleteness theorems, no proof of the independence of the continuum hypothesis and of the Axiom of Choice (hence, no 'non-Cantorian mathematics' would be possible), no adequate distinction between the theory's language and its metalanguage with all the metalogical consequences (for instance, about the concept of truth and all discussions around it), and so on. But there is still a question to be answered: should this technique be applied to all domains, to all sciences? That is, is it desirable, or should we take it as an ideal that every theory be formally axiomatized? Here we make a digression on this point and put a challenge, suggesting that perhaps the axiomatic technique is not something to be pursued in all domains. Let us see why by focusing on physics (we leave out human sciences in general, where the answer seems to be no most of the time, despite the advantages the AM may bring to these fields, as the works of Suppes have shown).

We shall focus on present-day physics of matter. Although we are not specialists in this field, we think that we can say something on this respect, at least from a philosophical and foundational points of view. What is relevant for us is that such physics poses important concerns on the axiomatic study of physics. By 'present-day' physics of matter we mean the Standard Model (SM) of particle physics. Roughly speaking, SM is composed by some sub-theories: quantum electrodynamics (QED), the theory of weak forces, and quantum chromodynamics (QCD). This theory unifies three of the four fundamental forces of nature, namely, the electromagnetic force, the weak force, and the strong force. Only gravitation is left out of the picture. The mathematical description is in terms of group theory; they are *gauge theories*, and we need to know their gauge groups. The first two sub-theories are unified as what is known as the *electroweak theory*, whose gauge group is termed $SU(2) \times U(1)$. It is not relevant for us here to explain the details, but only to be aware of the differences. The gauge group of QCD is $SU(3)$. The problem is to find a gauge group unifying all these groups. Some authors try to define the group $SU(3) \times SU(2) \times U(1)$, whose combination is the symmetry group $SU(5)$ [GlaGio.00]. The problem is that, until now, there was no agreement about what $SU(5)$ was, as Glashow tells us. In other words, until now there is no consistent way of joining the electroweak theory with QCD. They work as isolated pieces of a greater framework and are used depending on the subject the scientist is interested in. But things are not so easy even for these sub-theories. Let us follow some masters on the subject, Arthur Jaffe and David Gross (Nobel Prize winner 2004).

Gross suggests that QCD does not exist as a mathematical sound theory [Cao.99, p.164]. Jaffe suggests that this is a question of time and compares the issue with some results in mathematics, where there are results not known in a time, but that were put out on a well-established basis some time later. Gross doesn't disagree and claims that the search for a mathematically adequate basis is something that should be looked for. His point concerned present-day physics, and both of them agree that these fields are not yet well developed in regard to what concerns foundational aspects.

Thus we may conclude that, at least in the present moment, physical theories form a mosaic composed of different sub-theories, which are used depending on the field and on the particular problem being dealt with. Furthermore, sometimes these sub-theories may be even incompatible one with another, and even so, they constitute the mosaics of physics. Typical examples, besides those mentioned in the previous paragraph, are the quantum field theories (QFT—QED, QCD and the weak theory are quantum field theories) and general relativity, which comprises the fourth force, gravity. The attempts of joining the three forces with gravity conduce to Quantum Gravity, the Holy Grail of present-day physics. But Gross emphasizes also in his own paper in the book that an attempt to make sense to QFT (in general) is still required [Gro.99]. The problem is that we don't have this general theory today.

Well, should we search for such a unification in an axiomatic fashion?[11] From the foundational point of view, of course the answer is yes. The fact that we still do not have a fully developed unified theory may make attempts at axiomatization premature, but that is not a conclusive argument. Recall that one of the objections to the axiomatic method (by Hempel and others, see a discussion in Suppe [Sup.77, pp.110–5]) is that it really only predates a well-developed theory; the axiomatic method cannot work in a vacuum. In fact, it can only order a field of knowledge where there is something reasonably well-developed in order to be ordered. So the objection would go as follows: it is unreasonable to try to apply the axiomatic method to those areas of physics yet.

Our answer to those concerns comes from a balance between two extremes. We should obviously not force every field of knowledge into the form of the axiomatic method; they may well not be ready for that. Axiomatization is not a mark of scientific status per se. Not even the Logical Positivists were so rigid in their demands for formalization (see again Lutz [Lut.12]). On the other hand, to claim that no benefits result from the application of the axiomatic method would require a stubborn blindness to the evidences we have already pointed out of how useful it may be. Perhaps the best thing is to follow Halvorson [Hal.15, pp.2–3] and allow that science itself determines which are the fields of scientific knowledge that could profitably be axiomatized at a time. Scientific development itself seems to indicate that a field is ready for axiomatization and that it may be useful trying to axiomatize a theory. Obviously, this is as vague an answer as it gets, but we would not like our philosophical preferences to cause trouble for science.

4.5   DOES AXIOMATIZATION REALLY AXIOMATIZE?

The next question to be considered regarding the AM is about its results—something that was already advanced in the introduction. More specifically, the question is suppose we have an informal theory T, say Darwin's evolution theory classical physics before Newton (Galileo, Kepler, and others), but let us fix Darwin. Suppose now that we aim at axiomatizing his theory. This was done by some people, such as Mary Williams [Will.70][12] and many others since then. The question is does Williams's (or any other) system axiomatically capture Darwin's theory? It will all depend on how deep the idea of an axiomatic system capturing some informal theory goes. If such a capturing is understood as *complete identification between axiomatic and informal theory*, then perhaps the answer is 'no', or, at least, we cannot prove that. In our opinion, axiomatization and formalization create something that certainly *resembles* the original theory, but cannot be said *to be* that theory in the sense of identification. By axiomatizing T, we create T' and work *as if* T' were T; we get results in T' which are interpreted in terms of T and so on. This remark is to be taken into account in all that follows in this book.

Even informal axiomatics, as we have seen, which is most of the time advanced in order to *organize* a certain field of knowledge known in advance, and it is applied, as Lakatos and Hempel have suggested, only to 'mature' domains, surely leaves out much of the original theory, which has imprecise and wide borderlines. Thus, in elaborating an axiomatic version of a certain theory, the scientist has no way to assure us that the axiomatic system he or she has obtained really corresponds to the informal domain he or she had before. Let us make an analogy with a famous case. In recursion theory, there is an important result known as *Church thesis* (or *Church–Turing thesis*, referring to the American mathematician Alonso Church and the British Alan Turing) which says that all computable functions are recursive. In short, those functions defined on the natural numbers that are computable in the informal sense (we can effectively get the outcomes given some inputs) are formally computable, or recursive. Recursive functions have a precise definition, which does not concern us here. It is generally accepted that recursive functions are computable, for they were defined precisely to play this role. The thesis says the converse and *cannot be formally proven*, although some have tried. The reason is that we cannot use the resources of logic to compare a formal concept (that of recursive function) with an informal one (that of computable function). The point is that the informal counterpart lacks adequate precision to be compared with the concept analyzed on logical grounds. By the way, this is just one of the roles of the axiomatic method, namely, to produce a framework that can be analyzed in logical terms. The analogy is pertinent. In axiomatizing an informal field, such as Darwin's natural selection, we get an axiomatic or a formal theory that may resemble the original one, but due to the imprecision of the first, which in general has not had its basic concepts clearly stated or defined, the equivalence of the two must be accepted as a thesis, similar to Church's. In other words, we need to *postulate*, in general, grounded on the success of the axiomatic version in explaining the informal field that the two theories fit. But there cannot be a proof of this result.

As for another example, we present the usually mentioned fact that we can formulate Aristotle's theory of categorical syllogism in the language of first-order logic. It suffices to write the 'A' propositions (universal affirmatives) as $\forall x(S(x) \to P(x))$, where S and P are the subject and the predicate terms, respectively, the 'E' propositions (universal negations) as $\forall x(S(x) \to \neg P(x))$, the particular affirmatives ('I') as $\exists x(S(x) \wedge P(x))$, and the particular negations ('O') as $\exists x(S(x) \wedge \neg P(x))$. For postulates, we can use the inference rules derived from the first figure Barbara, Celari, Darii, and Ferio (or, more economically, it is enough to take Barbara and Celarent, as is well known), adequately translated to the logical notation. Have we succeed in 'reducing' syllogistic to first-order monadic predicate calculus? That is clearly doubtful. This *syllogistic theory*, inspired in Aristotle and in further developers, cannot be said to capture all the richness and complexity of the informal discussion given in the *Prior Analytics*, for instance, in distinguishing 'demonstrative premise' from 'dialectical premise'. The first one is either true or false and has been obtained from initial assumptions, while the dialectical one just poses a contradiction. In other words, the demonstrative premises are the parts of a contradiction, while the dialectical ones are just asking for the contradiction [Aris.89, 24a10, 24b10]. There are huge difficulties in translating this to our present-day logical language. If by a contradiction we understand the conjunction between a proposition and its negation (what depends on the meaning of these connectives, that is, of the logic involved), something like $\alpha \wedge \neg\alpha$, then both α and $\neg\alpha$ would be demonstrative premises, while the move to ask if this expression either is or not a contradiction would be considered as dialectic. Of course this discussion cannot be developed within such a poor formal theory.

But these remarks do not entail that the axiomatization of a certain field is not relevant, and let us insist on this point a little bit more. As a first point, we would like to recall an earlier discussion. Recall that we employ the axiomatic method either as a formal method or else as according to Suppes's suggestion, by devising a Suppes predicate. In both cases, we are axiomatizing a field of knowledge characterizing a theory for certain purposes. The fact that an axiomatized theory is a way of representing an informal theory (as French [Fre.15] insists) for philosophical purposes does not diminish its relevance. Philosophers have many intricate questions about science that require some way to represent theories. We would like to discuss what it means to say that a theory is true, or quasi-true, or that it is empirically adequate, and so on. How can that be done with an informal approach to theories? The axiomatic method helps us in discussing such issues, not only for specific cases where axiomatization is available but also *in abstracto*, by furnishing a general guideline of what a theory would look like when an axiomatization eventually is developed. So in this sense, a 'rational reconstruction' effected by the axiomatic method is certainly something to be sought by philosophers too.

As a second point, it is important once again to point to how useful the AM may be for scientific purposes. Suppose again that we are dealing with arithmetics. Without formalization, non-standard models would not become available for investigation. The same can be said of several other domains. Patrick Suppes emphasized the *heuristic* role of the axiomatic method. As an example, he mentions the axiomatization of the field of real numbers. As we have already mentioned, Suppes [Sups.83] says that Dedekind's and Cauchy's definitions of real numbers, respectively, by means of cuts and by equivalence classes of Cauchy sequences, are not easy to comprehend and deal with. On the other hand, the axioms of a complete ordered real field provide a useful mathematical device, and even the least-upper bound axiom can be understood without difficulty.[13] Kolmogorov's axioms for probability are the second example. According to Suppes, they provide a clear and intuitive foundation for the concept of probability. Thus a system of axioms is *heuristic*, according to Suppes, if they provide an intuitive and organized way of looking to the involved concepts, organizing and facilitating "our thinking about the subject, and in particular our ability to formulate, in an ordinary self-contained way, problems concerned with the phenomena governed by the theory and their solutions" (op.cit.). Non-heuristic axioms provide a "sophisticated mathematical foundation of a discipline, but [they are] formulated in such a way they prohibit natural and intuitive ways of thinking about problems, specially new problems in the discipline" (ibid.). As an example, we recall that he mentions Mackey's axiomatics for orthodox quantum mechanics. We have no space to discuss the details here, for which we recommend Suppes's paper.

The important point to be emphasized is, as we have suggested already, that without axiomatization, and more, without formalization, we would not be in a condition to notice that our room has some closed windows which, once opened, show us new landscapes and even new universes. This is true of the non-standard models of arithmetics, as we suggested, but also with non-standard models of real numbers, the rise of hypersets, obtained by dropping the axiom of foundation in set theory,[14] the existence of denumerable models of the first-order ZFC set theory (supposed consistent), Gödel's incompleteness theorems, Tarski's theorem about the undefinability of the notion of truth, and so on. It is not absurd for sure to suppose that the same new vistas may be opened with the axiomatic study of empirical sciences, once adequately formalized. In this sense, axiomatization is indeed welcome. We leave this question for the reader to think over as a challenge for the philosophical reflection, and advance now to the topic of axiomatization of scientific theories properly speaking.

NOTES

1. Textbooks in differential and integral calculus, analytic geometry, and many others are generally presented in an informal language, without axioms but only with definitions and theorems. But it is easy to note that these fields make part of a more detailed mathematical theories presented in an axiomatic way. Furthermore, it is known from logic that every system given by postulates can be transformed in another system described by nominal definitions only. The mentioned areas can be seen as described this way.

2. Although that would be useful. In fact, how can a mathematician speak of proofs—and usually make several proofs in her training and teaching—without knowing what a proof is?

3. Thus his celebrated formula $P_1 \to TT \to EE \to P_2$ [Pop.72, p.119/243].

4. Today we know that the constant k in $V - E + F = k$ is a *topological invariant* of the polyhedra of a kind. For instance, those where $k = 0$ are the topological equivalent to a torus and to a sphere if $k = 2$.

5. Even though we should acknowledge that the notion of 'most intuitive' is vague and relative.

6. Of course, in present-day mathematics, not all mathematical concepts can be developed within standard set theories; category theory is one such example. But such a claim depends on the set theory being considered; even category theory can be developed within *strong* set theories, for instance, those comprising universes. By adding to ZFC a postulate saying that a Grothendieck universe exists, we get a new theory so strong that category theory can be developed within it [Low.14]. Universes entail the existence of inaccessible cardinals.

7. Hamel said that the papers should not be published since the papers dealt only with "meager mechanics" of points, leaving aside, for instance, the problem of the continua — that is, the mechanics of the continuum, where deformations are considered.

8. Recall that in particle mechanics, the particles are taken to be pontual, without dimensions.

9. It is well known the claim made by Hilbert in his conference on the 23 Mathematical Problems that "[t]he mathematician will have also to take into account not only of those theories coming near to reality, but also, as in geometry, of all logically possible theories." [Hilb.76, p.15]

10. This proposition invites us to construct an equilateral triangle over a segment of line. The 'proof' assumes that two circles intersect, something that is quite 'obvious', but does not follow from the axioms.

11. This question does not ask for Quantum Gravity, but for an *axiomatic* quantum gravity.

12. The use of the axiomatic method in biology was initiated by John Woodger in the 1930s.
13. This axiom is essential to characterize the structure, for it provides its topological completeness.
14. Okay, you may say, these sets were already noted, for instance, by Mirimanov in 1917, but the importance of von Neumann's axiom of foundation cannot be ignored for the formulation of the intuitive notion of a set.

5 Axiomatization and Scientific Theories

Now that we have discussed axiomatization in general and the main features of the axiomatic method, it is time to put that method to work in the case of scientific theories. As discussed in a previous chapter, the fact that we are using formal tools and axiomatization does not collapse our approach directly in the traditional Received View. Rather, as we shall discuss later, we follow mainly the approach by Suppes and employ formal tools in the analysis of scientific theories wherever possible.

In this chapter, we shall proceed as follows. We begin by outlining what we call an 'external approach' to axiomatization — that is, an approach in which a theory is axiomatized along with all the mathematics required, in particular, set theory. We then evaluate rather briefly its inadequacies in the case of a treatment of the models of such theories. Basically, the main problem comes to the fact that by axiomatizing huge portions of set theory, an axiomatic system like that would have to have as models large parts of models of set theory as well! As we shall discuss very briefly, that is an impractical task for those willing to address philosophical problems concerning scientific theories (see our previous discussion).

The next step then consists of presenting two distinct approaches working inside set theory (recall that we have already mentioned that set theory would be our framework), so we may call those approaches 'internal approaches' to axiomatization. The first approach we shall present will be called 'da Costa–Chuaqui's approach. It is based on the work of da Costa and Chuaqui [CosChu.88] and da Costa and Rodrigues [Cos.07] and consists of a formalization of the concept of a *Suppes predicate*, by employing specific formal languages (the languages of the theory, or of their structures, as we shall see) built inside the language of set theory to frame the axioms. The second approach is the one by Suppes himself. It employs the language of set theory itself to develop the axiomatization of a theory (see [Sups.57] and [Sups.02]).

Both approaches differ in many respects. In particular, both are attempts to axiomatize a theory by selecting a class of models. However, their relations to models are very different, as we shall see, and the purposes for which each analysis is used may vary. We shall discuss the benefits and peculiarities of each approach, arguing in the end for a kind of pluralism in this aspect: the philosopher of science is free to employ the most appropriate analysis according to the needs of the day.

5.1 EXTERNAL AXIOMATIZATIONS

In the traditional debate between adherents of the so-called semantic approach against the Received View, a very common argument presented against the syntactical analysis of scientific theories concerns the nature of axiomatization required by the Received View. As we have already discussed, the folklore supports the claim that the axiomatization must proceed in first-order language with identity, by distinguishing a theoretical and an observational vocabulary and by providing an interpretation by the Correspondence Rules (cf. the standard exposition of the Received View in the introduction of [Sup.77]). Of course, that view has by now been somewhat demystified (see [Lut.12]), even though this kind of reading is still not widely known.

A typical argument against the plausibility of the Received View approach concerns the axiomatization of theories requiring large portions of mathematics, or even portions of set theory (this is a point advanced by Suppes [Sups.67, p.58]). In an axiomatization of geometry, for instance, one may need to use portions of set theory to deal with lines as sets of points. In more mathematical theories, such as probability theory, one may need the theory of real numbers. So axiomatization of those theories would require that we axiomatize portions of set theory and all the mathematics needed. Of course, one option is to use the set theory to develop the mathematics required, such as the theory of real numbers. Anyway, this kind of approach is said to be too laborious, requiring the unpractical axiomatization of all the relevant mathematics.

As we mentioned, it is now recognized that it would be unfair to attribute this view to the adherents of the Received View ([Lut.12] is very convincing about that). However, that is not enough to prevent one from axiomatizing theories according to that approach. In fact, Worrall [Wor.07, p.148] and Zahar [Zah.04, p.7] seem to promote this kind of axiomatization through their Ramsey approach to scientific theories. That is, their idea is that set theory is involved in the axiomatization of a scientific theory in the sense that one must either use set-theoretical notions in the axiomatization (Worrall) or else axiomatize set theory as part of the work of axiomatizing the theory itself (Zahar). So this is a live option for those adopting a syntactical approach to theories and a version of epistemic structural realism. Given that we are interested in the syntactical as well as in the semantic aspects of theories, we shall present this approach with some care and argue that it has serious shortcomings when it comes to dealing with its models. Later, we then move on to what we believe are more appropriate approaches.

According to the view being discussed, we can say that the axiomatic basis of a theory T encompasses three levels of postulates:

(i)   The logical postulates–in general, it is said that these encompass only those postulates of classical first-order logic with identity. But in the general setting, we could also use higher-order logics, infinitary languages or logics, or still other non-classical logics.

(ii) A group of 'mathematical' postulates–in general, postulates for first-order ZFC or, as Zahar prefers, NBG (von Neuman-Bernays-Gödel set theory, see [Men.97, chap.4]). Perhaps only a small part of ZFC is to be used, say Z^p, the 'pure' Zermelo set theory (but without *Urelemente*, although these entities may be interesting in empirical sciences to represent things that are not sets) plus the Axiom of Choice. The question of inquiring how much of set theory physical theories demand is still open and, of course, depends on the theory. But some mathematicians think that replacement axioms are not necessary and that only the denumerable form of the Axiom of Choice is enough.

(iii) The specific postulates–these postulates depend on the particular field being axiomatized. According to this scheme, these postulates would be sentences of the language of ZFC, perhaps enriched by additional concepts and terms referring to the supposed domain of investigation. In the case of the Received View, the specific vocabulary would still need to be separated into theoretical terms and observational terms.

The notion of deduction is that the underlying logic, in general, is in accordance (when no other logical symbols are introduced) with elementary (first-order) logic, but now in the deductions we may have axioms of ZFC as premises, so the logic is not 'exactly' elementary, as is necessary for the consideration of higher-order theories and languages typical of scientific theories. Indeed, in first-order set theory, the intended interpretation regards the individual variables as representing sets, viewed as collections of objects, which by their turn can also be sets. Since in extensional contexts we can read a set as being the extension of a property (the property shared just by the elements of the set), 'to belong to a set' stands for 'to have a property', so in quantifying over sets, we are quantifying both over individuals and properties. It is in this sense that set theory acts as a higher-order language, despite being (in most cases) formulated as a first-order theory.

For example, consider an axiomatization of classical particle mechanics. According to the approach here proposed, it would consist of an axiomatization for first-order logic, axioms for set theory, and, finally, in the third stage, axioms for classical particle mechanics. Notice that in formulating such axioms we may employ concepts of mathematics, such as vector spaces, real numbers and functions of real numbers. All those concepts may be developed inside set theory in the second level of the axiomatization. Surely this provides for a powerful 'mathematical basis' for a scientific theory by allowing all classical mathematics available. However, this approach to scientific theories also has its drawbacks.

What are the main difficulties with this kind of approach? Well, as we have seen earlier, ZFC (if consistent) cannot provide for its own models. So besides being an awkward approach for axiomatization, this form of presentation of theories is also unduly laborious for those willing to take into account the models of the theory: we will have to employ a stronger metatheory in which the models of ZFC (or whichever set theory we happen to be employing) are developed. So, for instance, whenever we are interested in particle mechanics or quantum mechanics, say, we will have to take into account not only the models of the specific axioms for those theories but also huge structures modeling the axioms of set theory. That goes beyond what any philosopher of science would be willing to take into account.

Perhaps some simpler approach that employs set theory but does not require a model of it could be more plausible. Let us check the alternatives. Before presenting the approaches, let us introduce the terminology that will guide us; it is common to both the Suppes approach as well as the da Costa–Chuaqui approach.

5.2   STRUCTURES

We shall be working within first-order ZFC set theory. In this section, we follow the approach presented in [Cos.07] but differently from the exposition there. We allow that the domain of the structures can be composed of more than one set and allow other simplifications that will be explained later.

Within ZFC, we can, at least in principle, construct particular structures for mathematics and for empirical sciences—sometimes we may need to strengthen ZFC, say with universes (which enable us to also deal with category theory, although we shall be restricted here to set-theoretical structures). As we shall see, the language of ZFC, termed $\mathcal{L}_\in$, taken here to be a first-order language, is so powerful that by using it, we construct languages that are not first-order and structures that are not order–1 structures (these definitions shall be introduced in what follows). To go into some detail, we need to introduce a few basic definitions we present in the next sections. A remark is in order here: we usually work with extensions by definitions of the language of set theory, comprising defined symbols such as $\subseteq$ (for set inclusion), $\aleph_0$, and so on. But we shall continue to speak of the language of set theory also for these extensions.

The general idea can be seen with an analogy with group theory. Starting with a non-empty set G, by using the set-theoretical operations, we obtain $\mathcal{P}(G \times G \times G)$, and then we may choose an element of this set (which is a set whose elements are collections of 3-tuples of elements of G), satisfying

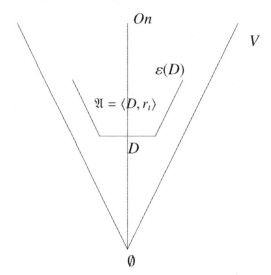

Figure 5.1   The well-founded 'model' $\mathcal{V} = \langle V, \in \rangle$ of ZFC, a structure $\mathfrak{A}$ built within ZFC, and the scale $\varepsilon(D)$ based on the domain D. For details, see the text

the required properties (the group axioms of course).[1] The corresponding structures (namely, the *groups*) are order–1 structures.

Summing up, what we are going to see is that a mathematical structure, such as those used in mathematics and in the empirical sciences, can be constructed within a set theory such as ZFC (with or without ur-elements). Figure 5.1 provides a schema of this situation, using 'pure' set theory (without ur-elements) with an axiom of foundation, as in standard first-order ZFC. It should be remarked that while the structure $\mathfrak{A}$ (to be defined soon) is a set, the whole universe V is not, as we have seen in the last chapter.

Our first definition is of the set $\mathbb{T}$ of types. These types are not to be confused with the types of Russell's Simple Type Theory. This set will be important also for the development of formal languages. We shall need to make explicit the da Costa–Chuaqui approach in the next sections. When we speak of the set of types, we always mean the set $\mathbb{T}$ given by the following definition:

Definition 5.2.1 (Types) *The set $\mathbb{T}$ of types is the least set satisfying the following conditions:*

(a) $0, 1, \ldots, n$, with $0 \leq n < \omega$, are types. These are the types of individuals, where ω is the first transfinite ordinal.

(b) if $a_1, \ldots, a_n \in \mathbb{T}$, then $\langle a_1, \ldots, a_n \rangle \in \mathbb{T}$, $1 \leq n < \omega$.

Thus, for instance, $0, \ldots, n-1$, $\langle 0 \rangle$, $\langle 0,0 \rangle$, $\langle 0,1 \rangle$, $\langle \langle 3 \rangle, 7 \rangle$, $\langle \langle 0 \rangle \rangle$ are types. Intuitively speaking, in this list we have types for the individuals of the basic sets, that is, the sets that form the domains of the structures, as we shall see soon. Thus the types just examplifyed stand for sets (or properties) of individuals whose elements have type 0 for binary relations on individuals of the same set for relations between the sets whose elements have types 0 and 1, respectively; for binary relations between sub-collections of the set whose individuals have type 3 and the set whose individuals have type 7; and, finally, for sets whose elements are subsets of the set whose elements have type 0.

Definition 5.2.2 (ORDER OF A TYPE) *The order of a type,* $\mathsf{Ord}(t)$, *is defined as follows:*

(a) $\mathsf{Ord}(k) = 0$, for $k = 0, \ldots, n$, with $0 \leq n < \omega$.
(b) $\mathsf{Ord}(\langle a_1, \ldots, a_n \rangle) = \max \{ \mathsf{Ord}(a_1), \ldots, \mathsf{Ord}(a_n) \} + 1$.

Thus, for instance, $\mathsf{Ord}(\langle 0 \rangle) = \mathsf{Ord}(\langle k_1, k_2 \rangle) = 1$, with $k_1, k_2 \in \{0, \ldots, n\}$, while $\mathsf{Ord}(\langle k_1, \langle k_2 \rangle \rangle) = 2$. *Relations* will be understood here extensionally (collections of n-tuples) and be of finite rank (that is, having finite elements only). *Unary* relations are sets.

Definition 5.2.3 (ORDER OF A RELATION) *The order of a relation is the order of its type.*

Thus binary relations of individuals are order-1 relations and so on. For instance, a binary relation on D is an element of $t(\langle k_1, k_2 \rangle) = \mathcal{P}(t(k_1) \times t(k_2)) = \mathcal{P}(D_{k_1} \times D_{k_2})$. According to definition 5.2.3, the type of binary relation on D_k is $\langle k, k \rangle$, as intuitively expected.

These definitions will be useful in our discussions to come, when we shall need to talk about the order of a structure and the order of a language. The following definitions help us to construct relations and properties based on $\{D_n\}$, a non-empty family of sets, each one with elements of a type, which will soon be called the domains of the structure. We write D_n for the nth member of that family. In the next definition, the usual set-theoretical operations of power set and Cartesian product are being used, denoted, respectively, by "$\times$" and "$\mathcal{P}$".

We shall introduce a function π as follows:

Definition 5.2.4 (SCALE BASED ON $\{D_N\}$) *Let* $\{D_n\} = \{D_1, D_2, \ldots, D_n\}$ *a family of non-empty sets whose elements have types* $t_1, \ldots, t_n$, *respectively. Then, we define a function* π, *called a scale based on* $\{D_n\}$ *with* $\mathbb{T}$ *as its domain, recursively, as follows:*

(a) $\pi(t_k) = D_k$ $(k = 1, \ldots, n)$.
(b) If $a_1, \ldots, a_n \in \mathbb{T}$, then $\pi(\langle a_1, \ldots, a_n \rangle) = \mathcal{P}(\pi(a_1) \times \ldots \times \pi(a_n))$.

(c) The scale based on $\{D_n\}$ is the union of the range of $\pi(\tau)$, with $\tau \in \mathbb{T}$ and it is denoted by $\varepsilon(\{D_n\})$.

Let us give an example. Since a binary operation on a set G (as the group operation) can be seen as a ternary relation on G, we can deal with it as follows. Having the scale $\varepsilon(G)$ (here the family of sets has just one element, the set G, and let t be its type), we just take an element of $\pi(\langle t, t, t \rangle) = \mathcal{P}(D \times D \times D)$ satisfying well-known conditions (the group postulates written in this 'relational' notation). In this sense, it can be shown that we can map Bourbaki's *echelon construction schema* [Bou.68, chap.4] within this schema.

Definition 5.2.5 *The cardinal K_{D_n} associated to $\varepsilon(\{D_n\})$ is defined as*

$$K_{D_n} = sup\left\{ \left|\bigcup_{k=1}^{n} D_k\right|, \left|\mathcal{P}\left(\bigcup_{k=1}^{n} D_k\right)\right|, \left|\mathcal{P}^2\left(\bigcup_{k=1}^{n} D_k\right)\right|, \dots \right\}.$$

Here $\left|\bigcup_{k=1}^{n} D_k\right|$ denotes the cardinal of the set $\bigcup_{k=1}^{n} D_k$.

Definition 5.2.6 (STRUCTURE) *A structure $\mathfrak{E}$ based on a family $\{D_n\}$ is a $n + 1$-tuple*

$$\mathfrak{E} = \langle D_1, \dots, D_n, R_\iota \rangle, \tag{5.1}$$

where $D_i \neq \emptyset$ and R_ι represents a sequence of relations of degree n belonging to $\varepsilon(\{D_n\})$. These relations are called the primitive elements of the structure.

Here R_ι is a sequence of elements of $\varepsilon(\{D_n\})$, and we suppose that the domain of this sequence is strictly less than K_{D_n}. We say that K_{D_n} is the cardinal associated with $\mathfrak{E}$ and that $\varepsilon(\{D_n\})$ is the scale associated with $\mathfrak{E}$.

As we said before, each element of $\varepsilon(\{D_n\})$ has a certain type, for it belongs to $\pi(t)$ for some $t \in \mathbb{T}$. Now, as we have seen, the order of a relation is defined as the order of its type. The order of $\mathfrak{E}$, denoted $\mathsf{Ord}(\mathfrak{E})$, is the order of the greatest of the types of the relations of the family R_ι, if there is one, and if there is no such relation, we put $\mathsf{Ord}(\mathfrak{E}) = \omega$. Usually, the relations of R_ι have as relata the elements of the basic sets, but more general structures enable that the relata can also be higher-order individuals, that is, sets of elements of D_i, sets of sets of these elements, and so on.

In the beginning of this section, we remarked that our presentation makes some simplifications on [Cos.07]. Here we depart from da Costa's original work in that we allow individuals and operations to occur in the structure, whereas da Costa reduced operations to relations and identified individuals with their unit sets. The main point of this change is to simplify the exposition, and from a mathematical point of view, the difference is purely a matter of convention.[2] So, in the definition of structures, the objects in the family R_ι may not only be relations, but operations as well — that is, relations satisfying the well-known functional condition, or even distinguished elements from the domain, which we take to be $0-ary$ operations. In these cases, we employ as usual the common notation for functions and objects.

Let us give some examples with due simplifications in the notation. For instance, a structure such as

$$\mathfrak{K} = \langle K, +, \cdot, 0, 1 \rangle \tag{5.2}$$

can be used for representing fields, where K is a non-empty set, $+$ and $\cdot$ are two binary operations on K, and $0,1 \in K$. Postulates for fields are added to this description. For vector spaces over a field $\mathfrak{K}$, we may write $\mathfrak{V} = \langle \mathcal{V}, K, +_V, \cdot_V, +_K, \cdot_K \rangle$, where $\mathcal{V}$ is the set of vectors K is a set, the domain of a field $+_V$ is a binary operation on V, called addition of vectors $\cdot_V$ is a function from $K \times V$ in V, the multiplication of vector by scalar $+_K$ is a binary operation on K, the addition of scalars and $\cdot_K$ a binary operation on K, the multiplication of scalars. All these concepts are subjected to well-known postulates. For simplicity, we usually omit the reference to field operations, leaving them presupposed in saying that $\mathfrak{K}$ is a field, and write simply

$$\mathfrak{V} = \langle V, K, +, \cdot \rangle. \tag{5.3}$$

More sophisticated structures suitable for representing theories in the empirical sciences, as we shall see in what follows, follow the same schema. For instance, as we shall see, a structure for classical particle mechanics is of the form

$$\mathfrak{M} = \langle P, T, m, \mathbf{s}, \mathbf{f}, \mathbf{g} \rangle,$$

where P is a non-empty set, T is an interval of the real number line, m is a function from P to the set of non-negative reals, and the remaining concepts are suitable functions we shall see again later in this chapter. Surely the reader acknowledges that, with some effort, this structure can be composed according to the aforementioned definitions.

5.3 LANGUAGES AND ALGEBRAS

We now present the approach to axiomatization based on the development of da Costa and Chuaqui [CosChu.88]. This work is an attempt to formalize the idea of the *Suppes predicate*, employing formal languages and characterizing a class of models by employing formal axioms. By a Suppes predicate we mean a set-theoretical predicate in the lines of Suppes, but, as we shall see, there are some differences in the definition and the way Suppes uses it. Now we shall discuss an approach to formal languages that will make use of the notion of structure just developed. Here languages will be seen as a special kind of structure: a free-algebra. Our aim is to develop some languages as the language of simple type theory, but before doing that, first we recall some useful concepts of Universal Algebra.

Definition 5.3.1 *Let R_i be a family of n-ary relations over sets whose elements have types $t_1, \ldots, t_m$, and let $D_1, \ldots, D_n$ be a family of sets whose elements have types $t_1, \ldots, t_n$, respectively. Thus the relations are relations over these sets by hypothesis. Let us consider the structure $\mathfrak{E} = \langle D_1, \ldots, D_n, R_i \rangle$. The similarity type, or signature, of this structure is a n-tuple $\langle a_1, \ldots, a_n \rangle$, where the a_i are the types of relations in the structure.*

For instance, a structure for group theory can be taken as $\mathcal{G} = \langle G, \circ, -, e \rangle$ (as we shall see below). Here $\circ$ is a binary operation over G (hence a ternary relation on G, $-$ is a mapping from G to G and $e \in G$. Hence the type of similarity of the structure is $\langle \langle 0, 0, 0 \rangle, \langle 0, 0 \rangle, 0 \rangle$.

We say that two structures have the same similarity type when the n-tuple $\langle a_1, \ldots, a_n \rangle$ of types is equal for both; that is, their relations have the same type and the family in the domain is composed of the same number of elements. Since a family is always ordered, the same type of relations always occur in the same order when the structures have the same similarity type. A structure for fields is of the similarity $\langle \langle 0, 0, 0 \rangle, \langle 0, 0, 0 \rangle, 0, 0 \rangle$, and so on.

Here we shall use a notion of *algebra* to mean mainly a structure composed by a set (the domain of the algebra) and a collection of operations and/or distinguished elements of the domain, according to the standard use in Universal Algebra [BarMac.75, p.2]. Thus, being $s = \langle a_1, \ldots, a_n \rangle$ as noted earlier, a n-tuple of types, we have:

Definition 5.3.2 *A s-algebra $\mathfrak{A}$ is a structure $\langle D_1, \ldots, D_n, R_i \rangle$ such that* Ord$(R_i) \leq 1$ *for each i, and $\mathfrak{A}$ has similarity type s.*

The restriction to order-1 or less serves to make each s-algebra an algebra in the usual sense — that is, a set with a family of operations and/or distinguished elements defined over this set and distinguished elements taken from the domain.

The definitions that follows will concentrate on structures having a sole domain for simplicity. The general case can be obtained by extending the mappings to all the sets of the domain, if there are many, by a procedure similar to one given by Bourbaki [Bou.68, chap.4]. More general structures, as those we have been mentioning, will appear again soon.

Definition 5.3.3 (Homomorphism between algebras) *Let $\mathfrak{A}, \mathfrak{B}$ be s-algebras with $\mathfrak{A} = \langle A, R_i \rangle$ and $\mathfrak{B} = \langle B, R_i' \rangle$. A homomorphism from $\mathfrak{A}$ into $\mathfrak{B}$ is a function $\varphi \colon A \mapsto B$ such that, for all R_i of the s-algebra $\mathfrak{A}$ whose type is $\langle a_1, \ldots, a_k \rangle$,*

$$\varphi(R_i(x_1, \ldots, x_k)) = R_i'(\varphi(x_1), \ldots, \varphi(x_k)).$$

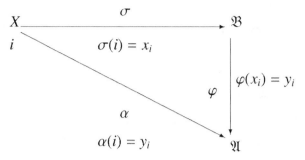

Figure 5.2   Free algebras: $\varphi \circ \sigma = \alpha$; that is, the diagram commute

Let us give an example using groups. Suppose two groups $\mathcal{G} = \langle G, \circ \rangle$ and $\mathcal{G}' = \langle G', \circ' \rangle$. A homomorphism from $\mathcal{G}$ into $\mathcal{G}'$ is a function $\varphi \colon G \mapsto G'$ such that, for every $a, b \in G$, we have

$$\varphi(a \circ b) = \varphi(a) \circ' \varphi(b).$$

A homomorphism between two vector spaces $\mathfrak{W}$ and $\mathfrak{W}'$ over the same field is a linear transformation from $\mathfrak{W}$ to $\mathfrak{W}'$.

Definition 5.3.4 (ISOMORPHISM) *Let $\mathfrak{A}$, $\mathfrak{B}$ be s-algebras and $\varphi \colon A \mapsto B$ a homomorphism. If φ is a bijection, we say that it is an isomorphism between the algebras.*

Definition 5.3.5 (FREE ALGEBRA) [3] *Let X be any set, let $\mathfrak{F}$ be a s-algebra with domain F, and let $\sigma \colon X \to F$ be a function. We say that $\langle F, \sigma \rangle$ (also, for simplicity, denoted by F) is a free s-algebra on the set X of free generators if, for every s-algebra $\mathfrak{A}$ and function $\alpha \colon X \to A$, there exists a unique homomorphism $\varphi \colon F \to A$ such that $\varphi \sigma = \alpha$.*

Figure 5.2 intends to clarify the definition using a simpler notation.

Theorem 5.3.1 *For any set X and any similarity type s, there exists a free s-algebra on X. This free s-algebra on X is unique up to isomorphism.*

We shall not prove this theorem here (but see [BarMac.75]). Now we pass to the development of languages properly, which will be considered as specific free s-algebras as presented earlier.

Let us suppose a simple case to exemplify how can we see what a language is from an algebraic point of view. In order to make the ideas clear, we shall consider an example in parallel by using the language of classical propositional logic, $\mathcal{L}_{\text{CPL}}$.

Thus suppose we have a set $X = \{x_i\}_{i \in I}$ (this set may be infinite) and another set f_λ whose elements will be denoted generically by f. We shall use for this set the same notation we used earlier to define an algebra. This second set plays the role of the operations of the algebra we shall define. In our sample case, $X = \{p_0, p_1, \ldots\}$ is the set of propositional variables, and $f_\lambda = \{\neg, \rightarrow\}$ are the propositional connectives (we could use any adequate set of connectives, of course). We still suppose that there is a mapping $\text{ar} : f_\lambda \mapsto \omega$, the arity function, which assigns a natural number to each element of f_λ, called its arity. For instance, $\text{ar}(\neg) = 1$ and $\text{ar}(\rightarrow) = 2$.

Now we define, by induction, a set $\mathcal{F}$ as follows:

(1) $F_0 := X \cup \{\langle f, x_1, x_2, \ldots, x_{\text{ar}(f)}\rangle : f \in f_\lambda \wedge x_i \in X\}$. This is the set of the *atomic formulas*. The $((\text{ar}(f) + 1))$-tuple $\langle f, x_1, x_2, \ldots, x_{\text{ar}(f)}\rangle$ is written $f x_1 x_2 \ldots x_{\text{ar}(f)}$. In our example, the atomic formulas are the propositional variables and the expressions of both forms $\neg p_i$ and $\rightarrow p_i p_j$. This last one may be abbreviated by $p_i \rightarrow p_j$.

(2) Now let $\alpha_i, i = 1, \ldots$ denote atomic formulas. Then the set of *complex formulas* is $F_1 = \{\langle f, \alpha_1, \alpha_2, \ldots, \alpha_{\text{ar}(f)}\rangle : f \in f_\lambda \wedge \alpha_i \in F_1\}$, and again the tuples are abbreviated by $f \alpha_1 \alpha_2 \ldots \alpha_{\text{ar}(f)}$. For instance, in our example we may have as complex formulas expressions such as $\neg\neg p_i, \rightarrow \neg p_1 \rightarrow p_2 \neg p_3$. The last one is abbreviated by $\neg p_1 \rightarrow (p_2 \rightarrow \neg p_3)$ according to a standard notation. In our case study, we can introduce other operational symbols by definition, such as the other standard connectives $\wedge$, $\vee$ and $\leftrightarrow$.

(4) In a general way, let $F_n := \{\langle f, \alpha_1, \alpha_2, \ldots, \alpha_{\text{ar}(f)}\rangle : f \in f_\lambda \wedge \alpha_i \in F_{n-1}\}$ and then

(5) $\mathcal{F} = \bigcup_{n \in \omega} F_n$.

We can prove that $\mathcal{A} = \langle \mathcal{F}, f_\lambda \rangle$ is a free algebra by adapting the proof given in [BarMac.75, p.5]. Thus, the language of the classical propositional calculus is a free algebra with $\aleph_0$ generators.

We call a free algebra constructed this way a *language*. Thus we can see how a language can be constructed in ZFC. The cases of first-order usual languages, higher-order languages, and even infinitary languages can be treated the same way, although they will demand more details. Anyway, they can be treated as being certain free algebras. Thus when we mention certain languages to speak of structures in the next section, they can be considered as constructed within ZFC.

5.4   LANGUAGES FOR SPEAKING OF STRUCTURES

Our goal is to speak of structures and of all the objects of a scale. To do that in an adequate way, we consider two basic infinitary languages, termed $\mathcal{L}^\omega_{\omega\omega}(R)$ (or simply $\mathcal{L}^\omega(R)$) and $\mathcal{L}^\omega_{\omega\kappa}(R)$[CosRod.07].

In general, an infinitary language $\mathcal{L}_{\mu\kappa}^{\eta}$, with $\kappa < \mu$ being infinite cardinals and $1 \leq \eta \leq \omega$, enables us to consider conjunctions and disjunctions of $n \leq \mu$ formulas and blocks of quantifiers with $m < \kappa$ many quantifiers. The superscript η indicates the order of the language (first-order, second-order, etc.). In both cases R is the set of the constants of the language. Thus in $\mathcal{L}_{\mu\omega}^{\omega}(R)$ ($\omega < \mu$) we may have infinitely many conjunctions and disjunctions of formulas, but blocks of quantifiers with finitely many quantifiers only. $\mathcal{L}_{\omega\omega}^{\omega}(R)$ is a higher-order language, suitable for type theory (higher-order logic). Standard first-order languages are of the kind $\mathcal{L}_{\omega\omega}^{1}$ then so is $\mathcal{L}_{\in}$.

Definition 5.4.1 (ORDER OF A LANGUAGE) *A language $\mathcal{L}_{\mu\kappa}^{n}$, with $1 \leq n < \omega$, is called a language of order n. A language $\mathcal{L}_{\mu\kappa}^{\omega}$ is said to be of order ω.*

A language of order n contains only types of order $t \leq n$ and quantification of variables of types having order $\leq n - 1$ [CosRod.07].

In order to exemplify how we can define a higher-order language using a first-order language (such as $\mathcal{L}_{\in}$), let us sketch the language $\mathcal{L}_{\omega_1\omega}^{\omega}(R)$, but we could consider whatever infinitary language $\mathcal{L}_{\mu\kappa}^{\omega}$, provided that the involved cardinals exist in ZFC (for instance, we couldn't use an inaccessible cardinal).[4]

The primitive symbols of $\mathcal{L}_{\omega_1\omega}^{\omega}(R)$ are the following ones:[5]

(i) Sentential connectives: $\neg$, $\wedge$, $\vee$, $\rightarrow$, $\bigwedge$, and $\bigvee$.
(ii) Quantifiers: $\forall$ and $\exists$.
(iii) For each type t, a family of variables of type t whose cardinal is ω.
(iv) Primitive relations: for any type t, a collection of constants of that type (possibly some of them may be empty). The collection of these constants forms the set R.
(v) Parentheses: left and right parentheses ('(' and ')') and comma (',').
(vi) Equality: $=_t$ of type $t = \langle t_1, t_2 \rangle$, with t_1 and t_2 of the same type.

Variables and constants of type t are *terms* of that type. If T is a term of type $\langle t_1, \ldots, t_n \rangle$ and $T_1, \ldots, T_n$ are terms of types $t_1, \ldots, t_n$, respectively, then $T(T_1, \ldots, T_n)$ is an *atomic formula*. If T_1 and T_2 are terms of the same type t, then $T_1 =_{\langle t_1, t_2 \rangle} T_2$ is an *atomic formula*. We shall write $T_1 = T_2$ for this last formula, leaving the type of the identity relation implicit. If α, β, α_i are formulas ($i = 1, \ldots$), then $\neg\alpha$, $\alpha \wedge \beta$, $\alpha \vee \beta$, $\alpha \rightarrow \beta$, $\bigwedge \alpha_i$, and $\bigvee \alpha_i$ are formulas. Then we are able to write formulas with denumerably many conjunctions and disjunctions. Furthermore, if X is a variable of type t, then $\forall X\alpha$ and $\exists X\alpha$ are also formulas (and only finite blocks of quantifiers are allowed). These are the only formulas of the language. The concepts of free and bound variables and other syntactic concepts can be introduced as usual.

Now let $\mathfrak{E} = \langle D, r_i \rangle$ be a structure, where $R_t \in R$ — that is, the primitive relations of the structure — are chosen among the constants of our language $\mathcal{L}_{\omega_1\omega}^{\omega}(R)$. Then, $\mathcal{L}_{\omega_1\omega}^{\omega}(R)$ can be taken as a language for $\mathfrak{E} = \langle D, r_i \rangle$, provided that $\kappa_D = \omega$ (recall that κ_D is the cardinal associated to $\mathfrak{E}$). Still working within (say) ZFC, we can define an interpretation of $\mathcal{L}^\omega_{\omega_1\omega}(R)$ in $\mathfrak{E} = \langle D, r_i \rangle$ in an obvious way so that what we mean by a sentence S of $\mathcal{L}^\omega_{\omega_1\omega}(R)$ (a formula without free variables) is *true* in such a structure in the Tarskian sense; that is,

$$\mathfrak{E} \models S. \tag{5.4}$$

In the same vein, we can define the notion of *validity*. A sentence S is valid, and we write

$$\models S,$$

if $\mathfrak{E} \models S$ for every structure $\mathfrak{E}$.

It is important to emphasize that we are describing the language $\mathcal{L}^\omega_{\omega_1\omega}(R)$ using the resources of some set theory such as ZFC. This way, we can speak of denumerable many variables, for instance, in a precise way. In this sense, any symbol of $\mathcal{L}^\omega_{\omega\kappa}(R)$, as we have remarked already, can be seen as a name for a set. Thus '(' (the left parenthesis), for example, names a set, as do all other symbols of the language.

5.4.1   The Language of a Structure

In this section, we relate both proposals made earlier: the theory of structures and the language of types seen as a free algebra. An interesting discussion in the philosophy of mathematics and foundations is whether there is a most natural language associated with a given structure. According to our approach, associated with every structure $\mathfrak{E}$ there will be a language that is the one in which we will talk about the elements of the structure, having as constant symbols exactly one symbol for each relation occurring in the structure, with both the symbol and the corresponding relation being of the same type. As we said before, given a structure $\mathfrak{E}$, to build one language for this structure we consider $\mathsf{Ord}(\mathfrak{E})$ and restrict ourselves to terms of this order or less. Also, the set X of generators of the free algebra will be restricted accordingly; that is, in building the language as a free algebra, the generators will be the atomic formulas built from the symbols available to us.

As a consequence of this discussion, the languages in which we can treat more adequately the elements of a structure are the languages $\pounds$ such that $\mathsf{Ord}(\mathfrak{E}) \leq \mathsf{Ord}(\pounds).$[6] So, for example, in first-order structures, we must use at least first-order languages. To consider a simple example, let's take group theory, which deals with groups (first-order structures). These structures, according to the approach adopted in this work, are most naturally treated by second-order languages, for we must talk about subgroups and quantify over subsets of the domain. That does not mean that it is impossible or not fruitful to use first-order languages; in fact, the work is commonly done in first-order language, as when we use set theory to develop group theory, or other mathematical theories such as well-ordered structures or Dedekind-complete fields, theories that are not first-order (see discussion in [Kun.09, pp. 89, 90]).

There is a heated debate on this topic in the philosophy and foundations of mathematics. Should we restrict ourselves to first-order languages or should we adopt higher-order languages? We shall not enter into the dispute here (for a defense of second-order logic in these contexts, see, for example, Shapiro in [Sha.91]). The general approach followed in our work suggests that a pluralist view can be fruitfully pursued: it is fruitful to explore higher-order languages as well as first-order languages; one must recognize that some kinds of structures are more naturally dealt with by using higher-order languages. In fact, the reader may even wonder whether the problem of a 'better' language for certain mathematical theories is a legitimate one (for agnosticism about this problem, see Hodges [Hod.01, pp. 71–3]).

Having a language with which we can talk about the elements of $\mathfrak{E}$, it is now simple to define for this language the notions of the structure modeling a sentence α of the language, that is, $\mathfrak{E} \models \alpha$, as well as other semantical notions, but we will not do that here.

So, taking the terminology introduced before this sub-section, let $\mathfrak{E} = \langle D, R_i \rangle$ be a structure, while $\mathsf{rng}(R_i)$ denote the range of R_i. Remember that R_i stands for a sequence of relations of the scale $\varepsilon(D)$; that is, it is a mapping from a finite ordinal into a collection of relations in the scale. Thus $\mathsf{rng}(R_i)$ stands for just the set of these relations. So $\mathcal{L}^\omega_{\omega\omega}(\mathsf{rng}(R_i)) \times (= \mathcal{L}^\omega(\mathsf{rng}(R_i)))$ is *the basic language of the structure* (it is not the only one, for other stronger languages encompassing it could be used instead). In this case, we can interpret a sentence containing constants in $\mathsf{rng}(R_i)$ in $\mathfrak{E} = \langle D, R_i \rangle$ and define the notion of truth for sentences of this language according to this structure in an obvious way.

Digression–For certain applications in science, sometimes it is better to consider *partial relations* as the primitive relations of a certain structure. A relation (say, a binary one) R on a set A is partial if there are situations where we cannot assert either aRb or $\neg(aRb)$ (see [CosFre.03] for all the philosophical discussion on this topic). In this case, the notion of truth is changed to *partial truth*, a concept that generalizes Tarski's approach and seems to be more adequate for empirical sciences. But we shall not touch this point here (but see [CosFre.03]).

5.5   DEFINABILITY AND EXPRESSIVE ELEMENTS

Now we wish to understand when an object of a scale $\varepsilon(D)$ is definable in a structure $\mathfrak{E} = \langle D, r_i \rangle$ by a formula of $\mathcal{L}^\omega(\mathsf{rng}(r_i))$ either when an element of the scale is expressible in the structure with respect to a sequence of objects of the scale.

Definition 5.5.1 (DEFINABILITY OF A RELATION) *Let R be a relation of type t =* $\langle t_1, \ldots, t_n \rangle$ *and* $\mathfrak{E} = \langle D, r_i \rangle$ *a structure. We say that R is definable in* $\mathfrak{E}$ *if there exists a formula* $F(x_1, x_2, \ldots, x_n)$ *of* $\mathcal{L}^\omega(\mathrm{rng}(r_i))$ *whose only free variables are* $x_1, \ldots, x_n$ *of types* $t_1, \ldots, t_n$, *respectively, such that in* $\mathcal{L}^\omega(\mathrm{rng}(r_i)) \cup \{R\}$, *the formula*

$$\forall x_1 \ldots \forall x_n (R(x_1, \ldots, x_n) \leftrightarrow F(x_1, x_2, \ldots, x_n))$$

is true in $\varepsilon(D)$.

For instance, for each type t we can define an identity relation $=_t$ as follows. Let Z be a variable of type $\langle t \rangle$ and then we can easily see that for suitable structures and scales, the following is true:

$$\exists! I_t \forall x \forall y (I_t(x, y) \leftrightarrow \forall Z(Z(x) \leftrightarrow Z(y))).$$

We may call I_t the identity of type t and write $x =_t y$ for $I_t(x,y)$, what intuitively means that identity is defined by Leibniz Law, as usual: just re-write the aforementioned definition as follows:

$$x =_t y := \forall Z(Z(x) \leftrightarrow Z(y)).$$

Usually, we suppress the index t and write just $x = y$, leaving the type implicit (= is of type $\langle t, t \rangle$, while x and y are both of type t). This kind of definability, which involves structures and scales, is called *semantic definability*, and goes back to Tarski.

Here is another example. Suppose the language $\mathcal{L}_\in$ of 'pure' set theory ZF. As is well known, this language has $\in$ as its only non-logical constant. If we attempt to define the subset relation $\subseteq$, we can do it in the extended language $\mathcal{L}_\in \cup \{\subseteq\}$ by showing that the formula

$$\forall x \forall y (x \subseteq y \leftrightarrow \forall z(z \in x \rightarrow z \in y))$$

is true in any structure built in ZF. In that which follows, we shall give some further examples.

Another important case is the next one, also involving a semantic definability.

Definition 5.5.2 (DEFINABILITY OF AN OBJECT) *Let us take* $\mathfrak{E} = \langle D, r_i \rangle$, $\varepsilon(D)$, *and* $\mathcal{L}^\omega(\mathrm{rng}(r_i))$ *as noted earlier. Given an object* $a \in \varepsilon(D)$ *of type t, we say that it is* $\mathcal{L}^\omega(\mathrm{rng}(r_i))$-*definable or definable in the strict sense in* $\mathfrak{E} = \langle D, r_i \rangle$ *if there is a formula F(x) in the only free variable x of type t such that*

$$\mathfrak{E} \models \forall x (x =_t a \leftrightarrow F(x)). \tag{5.5}$$

The case of the well-ordered set on the reals, mentioned earlier, shows that, taking into account the last definition, the least element of (0,1) cannot be definable by a formula. Let us consider a 'positive' example. Let $\mathfrak{N} = \langle \omega, +, \cdot, s, 0 \rangle$ be an order–1 structure for first-order arithmetics. In order to define a natural number (any one) we need just a finitary language, say $\mathcal{L}^{\omega}_{\omega\omega}(R)$ with $\mathfrak{R} = \{+, \cdot, 0, s\}$. Then it is easy to see that (in an obvious abbreviated notation)

$$\mathfrak{N} \models \forall x (x = n \leftrightarrow x = ss \ldots s(0)).$$

If we consider a suitable infinitary language, say $\mathcal{L}_{\omega_1\omega}$ (where we can admit denumerable infinite conjunctions and disjunctions), we can insert inside the parentheses the formula we abbreviate as follows:

$$x \in \omega \leftrightarrow x = 0 \lor x = 1 \lor \ldots , \tag{5.6}$$

which permits us to define not a particular natural number, but the notion of 'being a natural number'. An important remark: the expression (5.6) is not a formula strictly speaking (for the dots do not make part of the language), but abbreviates a formula of $\mathcal{L}_{\omega_1\omega}$.

An illustrative case is the following one. We know that within ZFC (supposed consistent), the set $\mathbb{R}$ of the reals is not denumerable. This means that we cannot find a mapping (a set in ZFC) that maps the reals onto the natural numbers. Thus, using standard denumerable languages, we do not have sufficient *names* for the reals—for instance, we can use each of the reals to name itself. But if we use a suitable infinitary language $\mathcal{L}_{\mu\kappa}$ (for suitable ordinals μ and κ) we can find a name for each real, so we can define all of them by the condition given in definition 5.5.2. This shows that definability and other related concepts depend on the employed language.

Here is another interesting case related to the aforementioned definitions. Using the Axiom of Choice, we can show that every set is well ordered (by the way, this statement is equivalent to the Axiom of Choice). For instance, the set ω of natural numbers is well ordered by the usual less-than relation $\leq$. We can define the usual order $\leq$ as follows:[7] $a \leq b := \exists c (b = a + c)$. The *usual* order relation, however, does not well order the set $\mathbb{Z}$ of the whole integers, for the subset $\{\ldots, -2, -1, 0\}$, for instance, has not a least element. But if we order $\mathbb{Z}$ by writing it as $\{0, -1, 1, -2, 2, \ldots\}$, then it is well ordered by this order, termed $\leq_1$, which can be defined by $a \leq_1 b := (|a| < |b|) \lor (|a| = |b| \land a \leq b)$, where $\leq$ is the usual order relation and $<$ is defined as $a < b := a \leq b \land a \neq b$.

Obviously, the 'usual' relation (defined on $\mathbb{R}$) does not well order the set of reals (for instance, an open set (a, b), with $a < b$, has not a least element). But can we find a well-order on such a set? According to the Axiom of Choice, we can assume that this order does exist, since its existence is consistent with ZFC (supposed consistent). The problem is that it can be proven that the ZFC axioms (plus the so-called generalized continuum hypothesis) are not sufficient to show that this order can be definable by a formula of its language.[8] In the same vein, we cannot define the least element of a certain subset of reals (say, the open interval $(0, 1)$) in the sense of definition 5.5.2. All of this shows that the notions of definability and expressibility, among others, depend on the language and on the theory we are assuming.

5.6   ON THE NEW SYMBOLS

In our formalization of ZFC, we have used individual constants to facilitate our mentioning of certain particular sets, for in doing that, we would have a denumerable set of 'names' for them. This is a matter of choice. Most authors don't use individual constants, but just individual variables. Since it is supposed that this move doesn't change the set of theorems, we regard the resulting theories as being the same, yet their languages differ. In this section, we shall make some remarks that we regard as philosophically important, which, although they are well known by the logician, may be not so well understood by the general philosopher interested in foundational issues. So let us suppose for a moment that our language $\mathcal{L}_\in$ has no individual constants. To avoid any confusion, we shall term it $\mathcal{L}_{\in^-}$.

Due to our new convention, the only non-logical symbol of $\mathcal{L}_{\in^-}$ is $\in$. So how can we refer to a particular relation R? (the same holds for any particular object such as a structure named $\mathfrak{A}$, for instance). This is a common practice in the mathematical discourse. Really, in geometry we usually say "Let A be a point in a line r", making use of individual letters for naming particular entities, the point, and the line. How can we explain this move of naming objects with constants that do not appear as primitive concepts of the employed languages? There are two answers to this challenge.

The first is that we simply extend the language $\mathcal{L}_{\in^-}$ with additional constants to name the objects we intend to make reference to. Thus we can extend $\mathcal{L}_{\in^-}$ with symbols of three kinds: (1) additional individual constants, (2) new predicate symbols, and (3) new operation symbols. In whatever situation, we must make sure that the so-called *Leśniewski's criteria* are being respected [Sups.57, chap.8], namely, the *Criterion of Eliminability* and the *Criterion of Non-Creativity*. The first says, in short, that the new symbols can be eliminated. That is, the formula S introducing the new symbol must be so that whenever a formula S_1 with the new symbol occurs, there is another formula S_2 without this symbol such that $S \rightarrow (S_1 \leftrightarrow S_2)$ is a theorem of the preceding theory (without the new symbol). The second criterion says that there is no formula T in which the new symbol does not occur such that $S \rightarrow T$ is derivable from the preceding theory, but T is not so derivable. In other words, no new theorem previously unproved and stated in terms of the primitive symbols and already defined symbols can be derived. In our case, we can add the desired symbols, say 'R' for the relation in the earlier definition of definability of a relation (see definition 5.5.1), once we grant that the Leśniewski's conditions hold (which of course we suppose here).

The other alternative is to work with $\mathcal{L}_{\in^-}$ proper and regard all other symbols as metalinguistic abbreviations. This is 'more economic', and we usually do it, for instance, when defining (in $\mathcal{L}_{\in^-}$) the concept of subset, posing that $A \subseteq B := \forall x(x \in A \rightarrow x \in B)$. The new symbol $\subseteq$ does not make part of the language $\mathcal{L}_{\in^-}$ but belongs to its metalanguage, and the expression $A \subseteq B$ simply abbreviates a sentence of $\mathcal{L}_{\in^-}$; namely, $\forall x(x \in A \rightarrow x \in B)$. We can understand this move as enabling us to use an *auxiliary constant* (say, 'R'), provided that the object that it will name exists (the proof of its existence is called *theorem of legitimation* by Bourkaki [Bou.68, p.32].[9] In the example, given two sets A and B, we realize that all elements of A are also elements of B, hence we are justified to write $A \subseteq B$ for expressing that (as a metalinguistic abbreviation). In logic, we usually express that by the so called *method of the auxiliary constant*, which may be formulated as follows. Let c be a constant that does not appear in the formulas A or B. Assume that we have proven that $\exists x A$ (the theorem of legitimation). If we have also proven that $A[^x_c] \vdash B$, where $A[^x_c]$ stands for the formula obtained by the substitution of c in any free occurrence of x, then $\vdash B$ as well.

But we need some care even here. Suppose we wish to refer to real numbers. We cannot name all of them in the standard denumerable language $\mathcal{L}_{\in^-}$. But in general, we can name a particular real number, say by calling it *zero*. But there are real numbers that cannot be named even this way, as we have seen before when we discussed the well-ordering of the reals. This poses an interesting question regarding empirical sciences. In constructing a mathematical model of a physical theory, we suppose we *represent* physical entities, say quantum objects and the properties and relations holding among them, in the mathematical framework we have chosen, say ZFC set theory. How can we ensure that this really makes sense? For instance, in some common interpretations of quantum mechanics (the Copenhagen interpretation), we really cannot (with any sense) make reference to quantum entities out of measurement. Out of measurement, quantum entities are no more than sets of potentialities or possible outcomes of measurement, to use Paul Davies's words in his introduction to Heisenberg's book [Hei.89, p.8]. Let us give an example we shall discuss next. Suppose we are considering the two electrons of a helium atom in the fundamental state.

The anti-symmetric wave function of the joint system can be written as

$$|\psi_{12}\rangle = \frac{1}{\sqrt{2}}(|\psi_1\rangle|\psi_2\rangle - |\psi_2\rangle|\psi_1\rangle), \tag{5.7}$$

where $|\psi_i\rangle$ ($i = 1, 2$) are the wave functions of the individual electrons. Notice that we need to label them by '1' and '2', for our languages are *objectual*—we speak of objects.[10] But we need to make this labeling not compelled with individuation, so we use (in this case) anti-symmetric functions, for the if $|\psi_{21}\rangle$

stands for the wave function of the system after a permutation of the electrons, then $\|\psi_{12}\rangle|^2 = \|\psi_{21}\rangle|^2$, that is, the relevant probabilities are the same. For now, what is relevant is that this function cannot be factored, giving a particular description of the electrons separately. Only after a measurement, say of the component of their spin in a given direction, does the wave function *collapse* either in $|\psi_1\rangle|\psi_2\rangle$ or in $|\psi_2\rangle|\psi_1\rangle$ and then indicate, say, that electron 1 has spin up in the chosen direction, while electron 2 has spin down in the same direction (or the other way around). But, before the measurement, nothing can be said of them in isolation. This implies that when we say, for instance, that there are two electrons, one *here* and another *there*, we are already supposing a measurement, thus begging the question concerning their individuation.

Situations such as this one are puzzling if we consider the underlying mathematics as the classical one.[11] Suppose we define the electron of a helium atom that has spin up in a given direction. According to definition 5.5.2., we need to find a formula $F(x)$ so that, if we denote that electron by a, we can prove that the following formula

$$\forall x(x = a \leftrightarrow F(x)) \tag{5.8}$$

is true in an adequate 'quantum structure'. But what would be taken to be $F(x)$?

Really, according to standard set theory (ZF), when we say that there is an electron here and another there, they are already distinct entities, and we are presupposing a kind of realism concerning these entities. In other words, in supposing *two* entities within ZF, we are really begging the question concerning their individuation.[12]

These remarks, we think, point to an interesting philosophical problem of studying the definability of physical 'objects' and relations, but this point will not be discussed here. Next, let us have a closer look at Suppes's and da Costa and Chuaqui's approaches.

5.7   DA COSTA AND CHUAQUI—SUPPES PREDICATES

In this section, we will return to the general concept of structure having as domain a family of sets. Given a structure and some language adequate for this structure, we now discuss how to formulate a Suppes predicate for that structure using that language, according to the directions suggested by N. da Costa and R. Chuaqui [CosChu.88]. First of all, we recall the definition of the similarity type of a structure $\mathfrak{E}$, which is a family of types that determines the kinds of relations present in the structure. According to this definition, two structures have the same similarity type if their type of relations form the same family and they have the same number of sets in their domain.

Now, given structures $\mathfrak{E} = \langle D_1, \ldots, D_n, R_i \rangle$ and $\mathfrak{G} = \langle E_1, \ldots, E_n, L_i \rangle$ of the same similarity type, we consider how to extend a given function $f : D_k \mapsto E_k$ for $k < n$ to a function mapping from $\varepsilon(\{D_n\})$ to $\varepsilon(\{E_n\})$.

Definition 5.7.1 *Given the function f as described earlier, we define:*

  (i) For the objects of type t, with $0 \le t < n$, let $f(D_k) = \{f(x) : x \in D_k\}$
  (ii) For $t \in \mathbb{T}$ such that $t = \langle a_0, \ldots, a_{n-1} \rangle$, and R the set of objects of type t, we have $f(R) = \mathcal{P}(f(D_{a_1}) \times f(D_{a_1}) \times \ldots \times f(D_{a_n}))$

This function maps objects of the type a in $\varepsilon(D_n)$ to objects of type a in $\varepsilon(E_n)$. The interesting case occurs when the following definition is verified.

Definition 5.7.2 *Given structures $\mathfrak{E} = \langle D_1, \ldots D_n, R_i \rangle$ and $\mathfrak{G} = \langle E_1, \ldots, E_n, L_i \rangle$ of the same similarity type s, and f a bijection from D_k to E_k with $0 \le k < n$, we say that the family $f' = f_s$ is an isomorphism between $\mathfrak{E}$ and $\mathfrak{G}$ when $f_t(R^t) = L^t$, where R^t and L^t means that R and L have type t.*

Definition 5.7.3 *A sentence Φ of the language appropriate for the structure $\mathfrak{E}$ is called transportable if for any structure $\mathfrak{G}$ isomorphic to $\mathfrak{E}$; that is,*

  $\mathfrak{E} \models \Phi \Longleftrightarrow \mathfrak{G} \models \Phi.$

Intuitively speaking, a transportable relation does not depend upon specific properties of the involved sets, but refers only to the way they enter in the relation, something that is given by the axioms. So a transportable relation cannot involve specific sets, but must speak of sets arbitrarily. So when in a relation it appears something like the empty set $\emptyset$, this is just a way of speaking, for this symbol can be substituted by the sentence $\exists y \forall x (x \notin y)$. We shall return to this remark later.

Definition 5.7.4 *A Suppes predicate is a formula $P(\mathfrak{E})$ of set theory that says $\mathfrak{E}$ is a structure of similarity type s satisfying Γ, a set of transportable sentences Φ of the language adequate for $\mathfrak{E}$.*

When $P(\mathfrak{E})$, that is, when $\mathfrak{E}$ satisfies P, we say that $\mathfrak{E}$ is a P-structure. According to da Costa and Chuaqui ([CosChu.88, p.104]), this definition captures the sense in which we can say that a theory is a class of models, precisely, the class of models that are P-structures for some adequate P.

We now consider some examples of Suppes's predicates for some theories.

5.7.1 A Suppes Predicate for Group Theory

Let G be a set; as defined earlier, we introduce the function t_G, or simply t, whose domain is the set $\mathbb{T}$ of types, to create the scale $\varepsilon(G)$. We suppose that the type of the elements of G is 0. Then we choose (1) a relation $\circ$ of type $\langle 0, 0, 0 \rangle$ — that is, $\circ \in \mathcal{P}(G \times G \times G)$; (2) a relation $-$ of the type $\langle 0, 0 \rangle$ — that is, $- \in \mathcal{P}(G \times G)$; and (3) an element e of type 0 — that is, $e \in G$. As one can check from the definitions, the order of each of the relations is 1 and the order of e is 0. Remembering that a n-ary function is a $n + 1$-ary relation, we have that the usual composition operation becomes a ternary relation, and the opposite relation becomes a binary relation.

The structure of groups is $\mathfrak{G} = \langle G, \circ, -, e \rangle$, and the order of this structure is the greatest order of its relations, so $ord(\mathfrak{G})$ is 1; that is, $\mathfrak{G}$ is a order-1 structure. The theory is not done yet, for we need to write down the postulates in order to give the set-theoretical predicate.

As defined earlier, the language for $\mathfrak{G}$ is a second-order language (note that the order of the language does not necessarily coincide with the order of the structure). The set T of terms is formed by a set of variables of type $t = 0$ or $t = 1$ and the set $\{\circ^{\langle 0,0,0 \rangle}, -^{\langle 0,0 \rangle}, e^0\}$ of constant symbols. So the set X of free generators is $X = \{T^t(t_0, \dots, t_{n-1}) : T^t \in T, \wedge t = \langle a_0, \dots, a_{n-1} \rangle \in \mathbb{T}\}$ $(t_k \in T_k)$. Based on the aforementioned theorem, there is free algebra on the set of free generators X, which is the language for the structure $\mathfrak{G}$.

With the language so developed we can write the usual axioms for group theory. Let's call them A1, A2, and A3, respectively, as follows:

A1 $\forall x \forall y \forall z (((x \circ y) \circ z) = (x \circ (y \circ z)))$,
A2 $\forall x \exists y (x - y = e)$,
A3 $\forall x (x \circ e = x)$.

Then a Suppes predicate for group theory can be written as follows:

$\mathcal{G}(X) \Longleftrightarrow \exists G \exists \circ \exists - \exists e (X = \langle G, \circ, -, e \rangle \wedge A1 \wedge A2 \wedge A3).$

The structures that satisfy this predicate are the models of the theory, namely, the groups. For more complex theories, of course we do not present the axiomatics of a theory this way but, as in standard mathematics, by describing the structure and listing the axioms. In the next example, we shall proceed this way.

5.7.2 A Suppes Predicate for Classical Particle Mechanics

First, we need to present some mathematical 'step' structures: the first one is the complete ordered field of real numbers $\mathfrak{R} = \langle \mathbb{R}, +, \cdot, 0, 1, < \rangle$. There is just one base set, the set $\mathbb{R}$ of real numbers. The objects of this set are of type 0. The operations are $+, \cdot, 0, 1, <$, which are of types $\langle 0, 0, 0 \rangle$, $\langle 0, 0, 0 \rangle$, 0, 0 and $\langle 0, 0 \rangle$, respectively (one must not confuse the symbol 0 of types with the element 0 of the field). As usual, these are the operations of addition, multiplication, the identity element of addition, the identity element of multiplication, and the relation of less than between real numbers. The language of the field of real numbers according to our approach is a second-order language, and the axioms for the field are well known, and we can easily see that they are transportable.

The next structure is the vector space $\mathcal{E} = \langle \mathcal{V}, \mathfrak{R}, +, \cdot, \mathbf{O} \rangle$ over the field of real numbers $\mathfrak{R}$. In this case, there are two base sets, the set $\mathcal{V}$ of vectors and the set $\mathbb{R}$ of real numbers. The objects of $\mathcal{V}$ are of type 1, while the real numbers are of type 0. Besides the field's operation, we have $+, \cdot, \mathbf{O}$, which are, respectively, of types $\langle 1, 1, 1 \rangle$, $\langle 0, 1, 1 \rangle$, 1. A Euclidean vector space is a vector space with the addition of an inner product and a vector product; we denote by $\langle \mathbf{x} | \mathbf{y} \rangle$ and by $[\mathbf{x}, \mathbf{y}]$ the inner product and the vectorial product of vectors $\mathbf{x}$ and $\mathbf{y}$, respectively. The first product gives a real number and the second one another vector, which are objects of types $\langle 1, 1, 0 \rangle$ and $\langle 1, 1, 1 \rangle$ respectively. The order of the language of vector spaces and that of the Euclidean vector space are 2 according to our approach — that is, a second-order language. The axioms for these structures are well known and clearly transportable.

Next, we present the real affine space, which is a vector real space with an addition domain A, whose elements are of type 2 and are called *points*. The new operation is the difference of points, which, for $p, q \in A$, is a vector denoted by $q - p$, whose type is $\langle 2, 2, 1 \rangle$; that is, the difference of two points gives us a vector. The new axiom for this concept is the statement that difference of points obeys the law of addition of points, which says that for $p, q, r \in A$, $q - p + r - q = r - p$. If the vector space is Euclidean, the affine space is a Euclidean space. In Euclidean space, we can define the distance between points $p, q \in A$ in the following way: $d(p, q) := \|q - p\| := \sqrt{\langle q - p | p - q \rangle}$.

Now we present a *Galilean space-time system*. We add to the four-dimensional affine space presented earlier in a new domain $\mathcal{V}_1$, which is a subset of $\mathcal{V}$ and operation $\mathbf{t}$ from A into $\mathbb{R}$ of type $\langle 2, 0 \rangle$ and relations of type $\langle 3, 3, 1 \rangle$ and $\langle 3, 3, 3 \rangle$ denoted by $\langle \cdot | \cdot \rangle$ and $[\cdot, \cdot]$ respectively, and $\mathbf{t}$ represents the measure of time. The two axioms following must be satisfied:

(a) $\mathcal{V}_1$ is a three-dimensional vector subspace of $\mathcal{V}$ and $\langle \cdot | \cdot \rangle$ and $[\cdot, \cdot]$ are its scalar product and vectorial product, respectively;

(b) $\mathbf{t}$ is a function from A to $\mathbb{R}$ such that for each $P \in A$ the set $\{Q : \mathbf{t}(Q) = \mathbf{t}(P)\}$ is a three-dimensional Euclidean space with vector space V_1. The affine space for $\mathbf{t}(P) = r$ is denoted by $A(r)$.

For a classical mechanical system, we need to add a new universe $\mathbf{P}$, the set of particles[13] and the set $\mathbb{N}$ of natural numbers to index the external forces. So the family of universes can be given by the sequence $\mathbb{R}$, $\mathcal{V}$, A,

V_1, **P**, and $\mathbb{N}$. The operations on these sets are those necessary to make $\mathbb{R}$, V, A, V_1 a *Galilean space-time system*, plus the following new relations:

(i) A function **a** of type $\langle 0, 2\rangle$, which gives the origin of a system of coordinates in each instant of time;

(ii) A function **s** of type $\langle 4, 0, 2\rangle$ for the position of a particle in each instant of time. We write $s_p(t)$ for this function;

(iii) A mass function *m* of type $\langle 4, 0\rangle$;

(iv) A force function **f** of type $\langle 4, 4, 0, 1\rangle$, which represents the internal forces;

(v) A force function **g** of type $\langle 4, 0, 5, 1\rangle$, which represents the external forces.

For the specific axioms of mechanics, we need notions of mathematical analysis such as derivatives and convergence of series. The field of real numbers must be completed with the corresponding operations of differentiation, integration and addition of series. Since differentiation, for example, takes functions of real numbers to functions of real numbers, the order of this operation will be 2. So the language needed to talk about this is structure is at least third-order language.

The kinematical axioms are:

(1) The range of **t** is an interval *I* of real numbers;

(2) **P** is a finite and non-empty set;

(3) **a** is a function from *I* to *A* such that for each $i \in I$ $a(i) \in A(i)$;

(4) **s** is a function from $\mathbf{P} \times I$ into *A* such that for each $p \in \mathbf{P}$ and $i \in I$ we have that $s_p(i) \in A(t)$;

(5) *m* is a function from **P** into $\mathbb{R}$;

(6) **f** is a function from $\mathbf{P} \times \mathbf{P} \times I$ into $V1$;

(7) **g** is a function from $\mathbf{P} \times I \times \mathbb{N}$ into V_1;

(8) For every $p \in \mathbf{P}$ and $i \in I$, the vector function $s_p(i) - a(i)$ is twice differentiable at i.

Dynamical Axioms:

(1) For $p \in \mathbf{P}$ $m(p)$ is a positive real number;

(2) For $p, q \in \mathbf{P}$ and $i \in I$ $f(p, q, i) = -(q, p, i)$;

(3) For $p, q \in \mathbf{P}$ and $i \in I$ $[s(p, i) - s(q, i), f(p, q, i) - f(q, p, i)] = O$;

(4) For $p \in \mathbf{P}$ and $i \in I$ the series $\Sigma_n(g(p, i, n))$ is absolutely convergent;

(5) For $p \in \mathbf{P}$ and $i \in I$ $m(p)D^2(s_p(i)) = \Sigma_{q \in \mathbf{P}} f(p, q, i) + \Sigma_n(g(p, i, n))$, where D^2 is the second derivative with respect to *i*.

These formulas are transportable in the sense defined previously.[14] The motivations for these formulas can be found in the works of Suppes cited in the bibliography. Other examples could (and perhaps should) be pursued, but these are enough to emphasize that in da Costa Chuaqui's approach, we need to consider the language of the structure and that we need also to axiomatize all the step structures necessary for the main theory. It leads to a Herculean effort, of course, practically similar to what was required by the approach suggested by the Received View. Now we shall turn to Suppes's own way of axiomatizing theories, and we shall see that it has some advantages due to its simplicity and power.

5.8   THE APPROACH OF SUPPES

The approach developed by Patrick Suppes in many of his works contrasts with the previous work by da Costa and Chuaqui. The technique of properly axiomatizing a class of structures inside set theory is described, for instance, in his [Sups.57]. He doesn't use a formal language specific to theory in study. In his work from the late 1960s [Sups.67], these kinds of approaches are distinguished as *intrinsic* characterization of a theory (when the axioms are formulated in a first-order language) and the *extrinsic* characterization of a theory (when the theory is axiomatized using the resources of informal set theory). Intrinsically, a theory is characterized when no appeal to any of its models is required, that is, by appealing just to that which is intrinsic to each model and, consequently, captured by a formal axiomatization. From an extrinsic point of view, a theory is characterized by describing directly the class of intended models as structures build in set theory.

Of course, in the context of the later mentioned work, Suppes distinguishes between presenting a class of structures through set-theoretical predicates in contrast to characterizing it as a formal elementary theory. He goes on and even calls our attention to the fact that it is not always clear whether a class of structures characterized extrinsically has an elementary axiomatization.[15] To provide an illustration, consider the class of well-orderings, that is, of structures $\mathfrak{W} = \langle D, R \rangle$ where R is a well-ordering on D. As is well known, there is no elementary axiomatization of this class.[16] However, we can give an extrinsic characterization of this class of structures (and, by leaving the restriction to elementary languages behind, the da Costa–Chuaqui approach presented earlier can also do it).

The distinction between intrinsic and extrinsic characterization and the benefits of exploring a scientific theory by looking at its models and not only at its syntactical characterization is a feature of Suppes's approach to the so-called semantic approach. In particular, his emphasis on models is closely related to his own approach to the relation between theory and experiment, which does not seem to be correctly described only in syntactical terms (as we have already discussed in our chapter 1). Anyway,

Suppes still needs a way to extrinsically characterize a class of models, and he does that by axiomatizing the theories inside set theory.

Suppes says he assumes intuitive set-theory, but we can continue to suppose that our discussion is conduced inside ZFC at an informal level, that is, without justifying every step of the required constructions. The first immediate advantage of starting with ZFC (or with some suitable set theory) is that one can employ all of the mathematics needed without explicitly having to axiomatize or listing its suppositions, for 'everything' we need is by hypothesis already done inside ZFC. Also, one leaves metamathematical considerations of usual axiomatizations, which proceed through the elaboration of formal systems and works in the known mathematical environment, simply by making use of the mathematics available. This last point was one of the attractions of the semantic approach over the Received View (see a full discussion in [Sup.77], but look also at [Lut.12] for a more realistic reading of the demands of the Received View), which proceeded through formal systems.

In general terms, a class of structures is axiomatized in set theory by providing for a list of axioms that a theory should satisfy. Those axioms are stated directly inside set theory using the vocabulary of set theory and further symbols that may be introduced by definition and that are specific to the theory. Let us see how we can define a set-theoretical predicate for groups. It is important to remark that now we do not need to describe all the step theories to be used, since we may assume that all of them can be given in the set theory being presupposed. Suppes gives various alternative formulations of this theory (mainly due to his concerns with the theory of definition and the search for an adequate and perspicuous form that conforms to some rigorous standards of definitions). To concentrate ourselves in one of the proposed ways of characterizing the theory, we can define a group through the following set-theoretical predicate, given informally first: a group is an algebra $\mathfrak{G} = \langle G, \circ \rangle$ where $G \neq \emptyset$ and $\circ$ is a binary operation on G, and ..., the dots being completed by the usual group axioms, written in the language of ZFC:

A1 $(\forall x, y, z \in G)(x \circ y) \circ z = x \circ (y \circ z)$
A2 There is an $e \in G$ such that for all $x \in G$, $x \circ e = e \circ x = x$
A3 $(\forall x \in G)(\exists x' \in G)(x \circ x' = x' \circ x = e)$

Now this is an example of a set-theoretical predicate for group theory and, as Suppes claims, it is entirely in conformity with the usual mathematical practice.

To be more precise, and using some terminology introduced before, we could specify the nature of the elements of the structure as well. So a group $\mathfrak{G}$ is an ordered pair $\langle G, \circ \rangle$ with $\circ \in t(\langle i, i, i \rangle)$ satisfying A1, A2, and A3 stated before. In symbols, the predicate $\mathfrak{G}$ is (or may be) the following one:

$$P(\mathfrak{G}) \leftrightarrow \exists G \exists \circ (\mathfrak{G} = \langle G, \circ \rangle \wedge G \neq \emptyset \wedge \circ$$
$$\in \mathcal{P}(G \times G \times G) \wedge A1 \wedge A2 \wedge A3). \tag{5.9}$$

Suppes calls a *group* any structure $\mathfrak{G} = \langle G, \circ \rangle$ that satisfies this predicate. Nothing specific is said about the satisfaction relation, or about the language in which the axioms are formulated. But, as we mentioned before, it is the enlarged language of the set theory (with new symbols such as $\mathfrak{G}$, $\circ$, etc.) that is being employed, and this is in conformity with the standard use in mathematics. How does one show that some structure is in the class characterized by the predicate so defined? For example, how does one show that $\langle \mathbb{Z}, + \rangle$ is a group? One simply derives, as theorems of ZFC, that the elements of $\mathbb{Z}$, along with the operation $+$, have the properties required for something to be a group. Then, $\langle \mathbb{Z}, + \rangle$ is said to *satisfy* the axioms for group theory and, so the structure is a *model* of group theory, or *it is a group*. We shall discuss precisely the relation of the axioms with the set-theoretical structures later.

Despite its great simplicity, the Suppes approach is very powerful indeed, and here we point to another of its advantages. While the standard first-order group axioms (to keep with our example) are enough to characterize *all* groups as being models of the axiomatics, in using set-theoretical predicates, we can axiomatize *certain* classes of models only. A typical example would be to axiomatize all groups except a certain specific case; for simplicity, let us suppose that we wish to leave out the additive group of the integers $\langle \mathbb{Z}, + \rangle$. We can do it by adding to the set-theoretical predicate given above a fourth condition $A4$ (another axiom), namely, a clause restricting the items that satisfy the predicate, say by requiring that those structures that satisfy it are groups but different from $\langle \mathbb{Z}, + \rangle$ (this can be easily expressed, for instance, by requiring in the set-theoretical predicate that $G \neq \mathbb{Z}$). This is most relevant for the semantic approach, which sometimes is said to be able to deal only with classes of intended models of theories [Sup.00, p.104].

It is worth noticing that an important difference between Suppes's and da Costa–Chuaqui's approaches is that in the former, the predicate does not need to be transportable, something considered crucial in the later, for the authors follow Bourbaki. The explicit reference to a specific set (in our sample case, $\mathbb{Z}$) shows that the predicate for the class of structures of groups except the additive group of the integers is not transportable.

For another example, one can take the Peano structures $\mathfrak{P} = \langle \omega, 0, \sigma \rangle$, with the usual Peano axioms in ZFC. Now the structure $2\mathfrak{P} = \langle 2\omega, 0, 2\sigma \rangle$, where 2ω is the set of even natural numbers and 2σ is the addition of 2, is also a model of these axioms. One can construct a Suppes predicate with the Peano postulates plus the requirement that the domain is different from 2ω, for example, and then this structure is not in the class of models of the set-theoretical postulate for arithmetics. This is possible because, in the language of ZFC, we can make reference to a particular element we leave outside the class of structures determined by the predicate. This fact is precisely what distinguishes the Suppes approach from da Costa–Chuaqui, and even from Bourbaki, let us insist once more. In the last two cases, the predicate (or the axioms in the case of Bourbaki) has two main parts: a *typification* and the axioms themselves. The typification is a formula, or a conjunction of formulas that specify the particular relations/operations/distinguished elements we are considering (for the formal definition, see [Bou.68, p.261]). Let us exemplify: in the case of groups, the typification is $\circ \in \mathcal{P}(G \times G \times G)$. In the case of vector spaces, we have $+ \in \mathcal{P}(V \times V \times V) \wedge \cdot \in \mathcal{P}(K \times V \times V)$. The axioms, in both cases must be *transportable* formulas; roughly speaking, this means that the definition of the formula does not depend upon any specific property of the construction made from the basic and auxiliary sets, but only refers to the way in which they enter in the relation through the axioms [Bou.68, loc.cit.]. In the Suppes approach, this restriction is not imposed; that is, the formulas needn't be transportable, and so we gain a wide range of extra possibilities.

For another example of these possibilities, let us consider a simple predicate for a structure composed by a set D and an operation σ picked from $t(\langle i, i \rangle)$ (that is, a binary operation on D) satisfying the conditions (axioms) (a) $\emptyset \in D$, (b) σ is injective, (c) $\emptyset$ does not belong to the image of σ, and (d) D is the least set satisfying these conditions. In this case, we are mentioning a particular item, $\emptyset$, and saying in the axioms that this chosen element must belong to the domain of the structure. Since we are making reference to a well-defined object (the set $\emptyset$), the formula that describes the set-theoretical predicate for the theory is not transportable. In this particular example, we still have that it is not possible to axiomatize $\mathfrak{M} = \langle D, \sigma \rangle$ in the standard way, say by fixing a specific vocabulary and making use of first-order logic (more on this later).

The fact that we can make such moves shows that set-theoretical predicates differ from Suppes's predicates as defined by da Costa and Chuaqui from Bourbaki's species of structures, although *some* predicates can fit the claims of transportability. The reason is that an axiomatics (say, in Bourbaki's style) should not make reference to specific objects such as the empty set. In fact, Bourbaki requires that the theory's postulates must be *transportable* formulas with respect to some typification, which teaches us how to deal with the added symbols of the structure. In other words, transportable formulas must be invariant by isomorphisms, and so they cannot make reference to specific objects [Bou.68, p.261–2]. These difficulties can be surpassed with some modifications and adaptations of our example, but they shall not concern us here. Anyway, in the Suppes-style axiomatization, we have free access to all mathematics that can be found in ZFC without constraints on the kinds of formulas that can be used to axiomatize a theory.

The previous discussion can be summarized in an interesting result. Recall that, when we restrict ourselves to first-order languages, in order to axiomatize a class of structures of order-1, we must give a set Γ of sentences such that all the structures in the class are models (in the sense of Tarski) of Γ and all models of Γ are in the class. A necessary and sufficient condition for a class of structures to be axiomatized by a set of sentences is that it must be closed by elementary equivalence and by formation of ultraproducts (for the definitions of those notions, see [Men.97]).[17]

Then for example, in this kind of axiomatization, when we deal with groups and the usual postulates for this theory written in a first-order language, we cannot have $\langle \mathbb{Z}, + \rangle$ out of the class of structures if $\langle 2\mathbb{Z}, 2+ \rangle$ is in this class (here $2\mathbb{Z}$ denotes the set of even integers and $2+$ addition of 2), for both are elementarily equivalent. The same holds for models of first-order Peano arithmetics, for the standard model $\mathcal{P} = \langle \omega, 0, s \rangle$ and $2\mathcal{P} = \langle 2\omega, 0, s' \rangle$, where 2ω is the set of even numbers and s' is the operation of adding 2. That is, we cannot have an axiomatization in a first-order language of one of them without having the other one as well, for they are elementarily equivalent. But, as we have seen, using the Suppes-style axiomatization, we can define classes of models that are order-1 structures but that are not closed by elementary equivalence. That is, Suppes's axiomatization is stronger than the usual one since it axiomatizes more classes of structures than it is possible to do with the usual procedure. To check this, we need only to consider the examples given earlier. So the examples show that there are classes of structures of order-1 axiomatizable by a Suppes predicate but not by a set of first-order set of sentences.

This result can be generalized to higher-order languages by following the same procedure (just take the class of all well-orderings except a particular well-ordering, for instance $\langle \omega, \in \rangle$).

For another example, compare now Suppes's axiomatization of particle mechanics with the one provided before by the da Costa and Chuaqui approach.

A system of particle mechanics is a 5-tuple[18]

$$\mathfrak{P} = \langle P, T, m, \mathbf{s}, \mathbf{f}, \mathbf{g} \rangle,$$

where P is a non-empty set (the 'particles'), T an interval in the set of real numbers (say expressing an interval of time), m is a function from P to $\mathbb{R}^+$ so that if $p \in P$, then $m(p)$ is the mass of p, $\mathbf{s}$ is a function from $P \times T$ in $\mathbb{R}^3$, so that $\mathbf{s}(p, t)$ is a vector expressing the position of the particle p at time t, $\mathbf{f}$ is a function with domain $P \times T \times I$, where I is a set of positive integers, so that $\mathbf{f}(p, t, i)$ is a vector representing the forces acting on p at t, and $\mathbf{g}$ is a vector representing the external forces acting on a particle at a given time t. All these concepts are subjected to the kinematical and dynamical axioms that follow:

Definition 5.8.1 *A Newtonian particle mechanics is a structure*

$$\mathfrak{P} = \langle P, T, m, s, f, g \rangle$$

satisfying the following axioms:
  Kinematical axioms

  P1 P is a finite non-empty set;
  P2 T is an interval of real numbers;
  P3   for $p \in P$, $\mathbf{s}_p$ is twice differentiable in T.

  Dynamical axioms
  P4   for $p \in P$, $m(p)$ is a positive real number;
  P5   for $p, q \in P$, $t \in T$, $\mathbf{f}(p, q, t) = -\mathbf{f}(q, p, t)$;
  P6   for $p, q \in P$, $t \in T$, $\mathbf{s}(p, t) \times \mathbf{f}(p, q, t) = -\mathbf{s}(q, t) \times \mathbf{f}(q, p, t)$;
  P7   for $p \in P$ and $t \in T$, $m(p)D^2\mathbf{s}_p(t) = \Sigma_{q \in P}\, \mathbf{f}(p, q, t) + \mathbf{g}(p, t)$.

As we see, it is easier to use this kind of procedure. A possible problem with this approach is that Suppes does not make reference to the specific set theory to be used in the axiomatization of a theory. We are free to use the theory we need to get the concepts we need. But there is also another kind of situation we shall describe now, which deals not specifically with the axiomatization of the theory, but with its interpretation, or *semantics*. Let us look at quantum mechanics to understand this problem.

5.8.1   An Axiomatization of Non-relativistic Quantum Mechanics

Let us consider a way to axiomatize non-relativistic quantum mechanics by means of a set-theoretical predicate following Suppes's approach. We shall be working informally, hoping that the reader understands that with a great boring effort, all could be done in terms of a 'legitimate' set-theoretical predicate in a set theory such as ZFC. We shall just present the relevant structures and the corresponding postulates, according to the standard mathematical practice. So we proceed as Suppes in his examples. Furthermore, we are not claiming that our axiomatics is adequate for all cases, since we restrict ourselves mainly to the finite dimensional case and to pure states only (mixtures are mentioned only in brief). So we shall be working in informal set theory, but if necessary, we may suppose a system such as ZFC. The Axiom of Choice seems to be essential here, mainly in the infinite dimensional case, for we need to grant that the vector spaces we use (Hilbert spaces) do have basis, and the general proof that a vector space has a basis depends on this axiom. The novelty in our approach is that we introduce a collection of physical objects (or systems) in the formalism, something that is absent in the standard approaches (really, in these approaches, the reference to physical objects is made only indirectly when we say—informally—that the vector state *represents* a quantum physical system; but this system never appears in the formalism). The definition is the one that follows, where, according to Suppes's approach, we presuppose some step theories given in set theory, such as the theory of Hilbert spaces, probability, differential equations, and all the involved mathematical apparatuses.

Definition 5.8.2 *A non-relativistic quantum mechanics structure is a tuple of the form*

$$\mathcal{Q} = \langle S, \{\mathcal{H}_i\}, \{\hat{A}_{ij}\}, \{\hat{U}_{ik}\}, \mathcal{B}(\mathbb{R}) \rangle, \text{ with } i \in I, j \in J, k \in K,$$

where

(i) S is a collection[19] whose elements are called *physical objects*, or *physical systems*.
(ii) $\{H_i\}$ is a collection of mathematical structures, namely, complex separable Hilbert spaces whose cardinality is defined in the particular application of the theory.
(iii) $\{\hat{A}_{ij}\}$ is a collection of self-adjunct (or Hermitian) operators over a particular Hilbert space H_i.
(iv) $\{U_{ik}\}$ is a collection of unitary operators over a particular Hilbert space H_i.
(v) $\mathcal{B}(\mathbb{R})$ is the collection of Borel sets over the set of real numbers.

The Hilbert space formalism, whose postulates we shall see in a moment, does not speak of space and time, something essential when we intend to apply it to the 'real world'. Next we comment on this important point.

Definition 5.8.3 *To each quantum system $s \in S$ we associate a 4-tuple of the form*

$$\sigma = \langle \mathbb{E}^4, \psi(\mathbf{x}, t), \Delta, P \rangle.$$

Here $\mathbb{E}^4$ is the Galilean spacetime;[20] each point is denoted by a 4-tuple $\langle x,$ $y, z, t \rangle$ where $\mathbf{x} = \langle x, y, z \rangle$ denotes the coordinates of the system and t is a parameter representing time, $\psi(\mathbf{x}, t)$ is a function over $\mathbb{E}^4$ called the *wave function* of the system, $\Delta \in \mathcal{B}(\mathbb{R})$ is a Borelian, and P is a function defined, for some i (determined by the physical system s), in $\mathcal{H}_i \times \{\hat{A}_{ij}\} \times \mathcal{B}(\mathbb{R})$ and assuming values in $[0, 1]$ so that the value $P(\psi, \hat{A}, \Delta) \in [0, 1]$ is the probability that the measurement of the observable A (represented by the self-adjunct operator $\hat{A}$) for the system in the state $\psi(\mathbf{x}, t)$ lies in the Borelian set Δ. We can see the relationship between the state vector and the wave function as follows. Let $\langle \mathbf{x}, t \rangle$ denote the location operation at time t. Then we put ψ

$(\mathbf{x}, t) = \langle(\mathbf{x}, t)|\psi\rangle$; that is, the wave function is described by the coefficients of the expansion of the state vector in the orthonormal basis of the position operator (a more precise description will be given soon).

These concepts are subjected to the following postulates:

Postulate 1 To each physical system $s \in S$ we associate a Hilbert space $\mathcal{H} \in \{\mathcal{H}_i\}$. Composite quantum systems are associated with complex Hilbert spaces that are the tensor product of the Hilbert spaces for each system, as usual.

Postulate 2 The one-dimensional subspaces of $\mathcal{H}$ denote the *states* the system may be in. These spaces are called *rays* by the physicists. To simplify the notation, usually they are represented by unitary vectors ψ (or by $|\psi\rangle$ in Dirac's notation) that generate these spaces. Hence $c\psi$, for $c \in \mathbb{C}$, represents the same state as ψ, as does $\psi.c$ (this is a typical physicist's abuse of notation).[21] These vectors are said to represent *pure states* of the system. It is also postulated that linear combinations of pure states, that is, vectors of the form $\psi = \Sigma_n a_n\psi_n$, for $a_n \in \mathbb{C}$ (the linear combination may comprise any finite number of vectors) called *superpositions* by the physicist, also denote pure states. This assumption is called *the Superposition Postulate*.

Postulate 3 To each observable (physical quantity that can be measured) A, we associate a self-adjoint operator $\hat{A} \in \{\hat{A}_{ij}\}$.

Postulate 4 The possible values of the measurement of observable A for the system s in state ψ lie in the spectrum (the set of eigenvalues) of the associated operator $\hat{A}$. This was called *the Quantization Algorithm* by M. Redhead [Red.87, p.5].

Postulate 5 Here we have the *Born Rule*. Given a system s, which is associated with a 4-tuple σ according to the aforementioned definition, let A be an observable to be measured on the system in state ψ. First we take the Hilbert space of the states of the system, $\mathcal{H}$. Now, let $\{\alpha_n\}$ be an orthonormal basis for $\mathcal{H}$ formed by eigenvectors of $\hat{A}$ (something that is possible to assume, since $\hat{A}$ is diagonalizable) so that there are complex numbers c_n such that $\psi = \Sigma_n c_n\alpha_n$, with $\Sigma_n|c_n|^2 = 1$. We know that the c_n are the *Fourier coefficients* $c_n = \langle\alpha_n|\psi\rangle$, where $\langle\cdot|\cdot\rangle$ is the inner product. Let us denote the eigenvalues associated with the vectors α_n by a_n; that is, $\hat{A}\alpha_n = a_n\alpha_n$. Then we have the *Statistical Algorithm* [Red.87, p.8]: the probability that the measurement of observable A gives the value a_n when the system in the state ψ is

$$\mathsf{Prob}_A^\psi(a_n) = |c_n|^2 = |\langle\alpha_n|\psi\rangle|^2$$

for the non-degenerate state (that is, all eigenvalues of $\hat{A}$ are distinct; when the operator is degenerate, the probability is obtained by summing the $|c_j|^2$ for all α's associated with the same eigenvalue).[22]

Other possible states that are not pure are called *mixtures*. They can be briefly described as follows by means of *statistical operators* (or statistical matrices) [Red.87, pp.15–16]. We assign probabilities w_k to a set of pure states $\{\beta_k\}$ in which the system may be found so that we have a *statistical ensemble* of several quantum (possible) states. Let P_{β_p} denote the projection operator whose range is the unitary sub-space generated by β_p. Then the statistical operator for the system becomes

$$\rho = \sum_k w_k P_{\beta_k},$$

and the *expectation value* of an observable A is given in terms of its Hermitean associated operator by

$$\langle A \rangle := \mathsf{Tr}(\rho \cdot \hat{A}),$$

where Tr is the trace function.

Postulate 6 Let us call this one *the Dynamic Postulate*. It says that if the system is in the instant t_0 in state $\psi(t_0)$—here the notation is adapted in order to consider the state as depending on time—then in a distinct time t the system evolves to the state $\psi(t)$ according to the Schrödinger equation [Pen.05, p.536], [Red.87, p.12]

$$\psi(t) = \hat{U}(t)\psi(t_0),$$

where $\hat{U}$ is a unitary operator.

Postulate 7 This is *the Collapse Postulate*. It says that immediately after the measurement of observable A for the system in state $\psi = \sum_n c_n \alpha_n$ gives the value $|c_n|^2 = |\langle \alpha_n | \psi \rangle|^2$, the system enters in the state described by the corresponding eigenvector α_n.

We shall now offer some hints about how to connect the aforementioned structure with possible applications.

Let us suppose that the state of a physical system is given by the vector ψ in a suitable Hilbert space. In order to represent observables such as position and momentum, we need infinitely many dimensional Hilbert spaces. So suppose a particle is moving in a line and let X be the position observable whose Hermitian corresponding operator is $\hat{X}$. If $\{\psi_x\}$ is an orthonormal basis for the Hilbert space formed by eigenvectors of $\hat{X}$, with λ_i being the corresponding eigenvalues, then $\hat{X}\psi_i = \lambda_i\psi_i$. Thus the state vector can be expanded in this bases, that is, we may write

$$\psi = \int \langle \psi_x | \psi \rangle \psi_x dx,$$

and the Fourier coefficients in this expansion enable us to define, or form the wave functions

$$\psi(x) := \langle \psi_x | \psi \rangle,$$

which are the functions $\psi(\mathbf{x}, t) = \langle (\mathbf{x}, t) | \psi \rangle$ described earlier with another notation. If we assume that ψ is normalized, that is $\| \psi \| = 1$, then we have

$$\int \psi(x)^* \psi(x) dx = 1,$$

where the star stands for complex conjugation. So $\psi(x)$ must be square integrable, that is, an element of the Hilbert space $\mathcal{L}_2$ of the square integrable complex functions.[23] But, in general, we need also not-square integrable functions, so we need to work in an expanded, or *rigged* Hilbert space, but we shall not enter this discussion here [Mad.05].

Of course, all non-relativistic quantum mechanics can be developed from this axiomatic basis.

One of the interesting questions that can be put here is the following one. Since quantum objects may be indiscernible, sharing all their characteristics (for instance, bosons in the same quantum state), should S be taken to be a set? We remember that in standard (extensional) mathematics a set is a collection of *distinct* objects.[24] We will not address this question in full here, but just to call the attention of the reader to the importance of considering the mathematical apparatus where the structures are developed. So let us give two examples.

The first one concerns semantics. The mathematician Yuri Manin said that quantum mechanics (better, its language) has no semantics. And more, he suggested that quantum mechanics has not even a language, a proper language, making use of a fragment of the language of standard functional analysis (the theory of Hilbert spaces in most presentations) [Man.10]. But, given the axiomatics presented earlier, we may say that there is a reasonable language involved, and we as philosophers or as logicians would like to know whether it is possible to ascribe to it a reasonable semantics in the standard sense. In order to do it, first we need a *domain of discourse*, let us call it D. But suppose, as it is common in many quantum cases, that the quantum objects being considered are *indiscernible* (or *identical* in the physicist's jargon, which is an abuse of language), say a collection of identical bosons in the same quantum state. In this situation, they have all properties in common, even spatio-temporal ones, if these notions make sense at all. In this case, we would not be able to discern the systems *by any means*. But if we are using ZFC as our metatheory, D is a set, hence its elements are always distinguishable, even if only in principle; if D is finite, they can be effectively discerned. This is, of course, contrary to the idea of the indiscernibility of the quanta. So we would be in need of a different metamathematical framework to elaborate our semantics. Such a theory, as we have seen already, exists and is called *quasi-set theory*, where we may have collections of objects that may be completely indiscernible! For details, see [AreKra.14], [DomHolKra.08], [FreKra.06], and [KraAre.15].

The second example concerns *names* of quantum entities. For the same reasons posed earlier, in certain situations, we should have no way to name certain quantum systems, even if they are fermions, which obey Pauli's Exclusion Principle. In fact, suppose that we name Peter and Paul the two electrons of a helium atom in its fundamental state (less energy). Supposing that we can speak of the electrons in this situation,[25] they are in a superposition and cannot be discerned. But let us suppose we wish to measure the spin of the electrons in a given direction, say in the z-direction. Thus we know from quantum mechanics that one of them will have spin UP in the chosen direction while the other one will have spin DOWN in the same direction. So they do not have *all* the same properties in common, saving Pauli's principle.[26] But which one has spin UP? Peter, Paul? There is no way to know the answer! If we admit that something is lacking, we need to assume that there are *hidden variables* of some kind and not all physicists are sympathetic to this idea. Thus, apparently, this is not a simple epistemological question in which we are not *able* to identify each of the two electrons: it is beyond doubt that the names have no meaning in this world. They are just *mock names*, which have no significance as proper names have in standard semantics.[27]

As we see, there are good reasons for looking for an alternative metamathematical framework, at least in the case of quantum mechanics. For further details on this point, see the aforementioned suggested references.

NOTES

1. The binary group operation ∘ is a certain function from $G \times G$ in G or, what is the same, it is a certain ternary relation on G, that is, a collection of triples $\langle a, b, c \rangle$ of elements of G, with $c = a \circ b$.
2. In fact, to allow individuals as we are doing or to identify them with their unit sets are both commitments to individuals anyway. If one is wishing to avoid individuals for structuralist reasons, then maybe ZFC is not the best framework to work with, since individuals are always given in the domain of the structure. Interesting as it is, we shall not concern ourselves with structuralism in this book.
3. Most texts on Universal Algebra speak of *absolutely free algebras*. We shall omit the term 'absolutely' from now on.
4. As we have seen in the previous chapter, there are cardinals whose existence cannot be proved in ZFC, provided this theory is consistent. Inaccessible cardinals belong to this class.

5. Of course we could use the schema of free algebras to characterize this language, but this would demand a lot of artificiality and will not conduce to anything relevant. The important thing is to acknowledge that the languages we will consider can be treated as free algebras.

6. We are following here also some suggestions made by Newton da Costa through personal communications in seminars.

7. Of course this definition can be conformed to definition 5.5.1.

8. The interested reader can check theorem 4.11 of S. Feferman's paper and the remark on p.342, just after the proof of the theorem; see http://matwbn.icm.edu.pl/ksiazki/fm/fm56/fm56129.pdf.

9. This is essentially what Mendelson calls 'Rule C'; see [Men.97, p.81].

10. Toraldo di Francia says that *objectuation* is a primitive act of our mind; see [Tor.81, p.222].

11. We shall return to this point at the end of this chapter.

12. This is, in our view, the main difficulty involved in attempts made by Muller and Saunders [MulSau.08] and by Muller and Seevinck [MulSee.09] to discern quantum entities; by assuming ZF as their mathematical framework, they are already assuming that the represented entities are either identical (that is, they are the very same entity) or that they are distinguishable, and this is not a characteristic of quantum objects proper, but of the mathematics they have employed.

13. It is important to realize for what follows that in this case **P** can be taken as a *set* of a set theory such as ZFC. The quantum mechanical case will be mentioned later, and in this situation, sometimes the domain of entities will not be not a set, as we shall see.

14. In fact, even in the case of the set $\mathbb{R}$ of real numbers, the general situation can be described taking any completed ordered field, so the axiomatics does not speak of any *particular* set.

15. Note that in the da Costa–Chuaqui approach we just presented, no restriction to elementary languages is required.

16. Perhaps it would be interesting to review some concepts from logic here. Let L be a first-order language and T a theory in L. A model M of T is an L-structure that satisfies the postulates of T. Let us call $M(T)$ the class (in general not a set) of all L-structures that are models of T. Let K be a class of structures. We say that K is elementary, or finitely axiomatizable, if there is a finite (a finite number of axioms) T such that $K = M(T)$.

17. We thank Antonio M. N. Coelho and Newton C. A da Costa for calling our attention to these points and for discussions on these themes.

18. Here we take a simplified version of this structure; see Suppes [Sups.57], chap. 12. In [Sups.02], Suppes provides an alternative formulation.

19. Next we shall question whether this collection can be considered a *set* of ZFC.

20. For details, which do not interest us here, see [Pen.05, chap.17].

21. The reason for this to be an abuse of notation is easy to explain. The operation of multiplication of a vector by a scalar (a complex number) is defined to be *multiplication to the left*; that is, it is a function from $\mathbb{C} \times \mathcal{H}$ to $\mathcal{H}$. If we want that this also represents a function from $\mathcal{H} \times \mathbb{C}$ to $\mathcal{H}$, we need to say it explicitly.

22. For details, see [Red.87, p.8].

23. Roughly, this is the space of the complex functions f such that $\int_{\infty}^{\infty} |f(x)|^2 dx < \infty$.

24. Cantor's 'definition' of the concept of set is well known: sets are collections of distinct objects of our intuition or of our thoughts [Can.55, p.85].

25. Thus we need to adopt an interpretation that is not 'instrumentalist', such as Bohr's, to whom quantum mechanics just provides us probabilities of measurements.
26. Electrons are fermions and cannot have all the same quantum properties in common. This is what Pauli's principle says, roughly.
27. This 'mock individuality' was an idea advanced by Toraldo di Francia; see [FreKra.06, p.361] also for references.

6   Models and Scientific Theories

In this chapter, we shall return to issues of a foundational-philosophical nature, where we address important questions about the nature of scientific theories in light of the developments made in the previous chapters. Questions concerning the nature of scientific theories are a constant presence in the philosophy of science, and as we saw in chapter 1, the debate on the precise nature of the semantic approach has recently reappeared. Our main aim is to explore the consequences that the specific framing of theories as exposed in the previous chapter may have when we deal with models, languages, and the general structure of scientific theories. The investigation will proceed at the metalevel.

As we have seen in the previous chapters, models are central to the semantic approach to scientific theories in both Suppes's and in da Costa and Chuaqui's views. But the word 'model' has distinct meanings in these two approaches. Here we recall that, just as we have done in previous chapters, we are following Patrick Suppes, according to whom most of the distinct senses of the word 'model' may be profitably studied as set-theoretical structures (on the plurality of meanings of the word 'model', among them *iconic* models, *analogy* models, *logical* models, and others, see the discussion in [Sups.02, §2.1]). As we have remarked in chapter 1, the current debate centers around the issue of whether a set theoretical structure must necessarily comprise an interpretation of a language or whether it can be merely a kind of 'language-free' structure. As we have seen, mostly as a result of the investigations by Lutz [Lut.15], the distinction between both approaches seems to be practically none. In this chapter, among other things, we shall see how da Costa and Chuaqui's and Suppes's approaches feature in these debates.

6.1   AGAIN ON THE ORDER OF LANGUAGES AND STRUCTURES

As a first important remark (which was already highlighted in chapter 1), it is usually claimed that the semantic or the model-theoretical approach to scientific theories was inspired somehow in Tarski's work of the 1950s, when he started the development of (standard) Model Theory, a branch of mathematical logic that deals mainly with the relations between *first-order* formal languages and set-theoretical structures that interpret them. So, by taking a glimpse in the relevant literature, we find that most philosophers think that the models of scientific theories are models of first-order languages in the Tarskian sense, which, as we shall see soon, is a simplification of the issue, given that in general the languages of most scientific theories are not first-order languages and the structures are not (as we shall call them) *order-1* structures. So let us begin by liberating the whole discussion from the limitation to first-order languages and structures that are adequate exclusively to such languages.

In the previous chapter, we have defined rigorously the order of a formal language. As it is common in most cases, these languages are *finitary languages*, comprising only formulas of finite length and, in particular, encompassing finite families of conjunctions and disjunctions, as well as finite strings of quantifiers. So we can characterize them with the general notation $\mathcal{L}^n_{\omega\omega}$, where n indicates the order of the language, $n \in \{1, 2, 3, \ldots\}$. If n is not a natural number, we say that the order of the language is ω. When $n = 1$, we speak of *first-order* languages, of *second-order languages* when $n = 2$, and so on. The structures where these languages are interpreted are *order-1* structures, and this means that we quantify only over individuals of the domain(s) of the structure, but neither over sets of such individuals (i.e. their properties) nor over elements of higher-order, such as sets of sets of individuals, and so on. Second-order languages admit predicate variables, which are interpreted (extensionally) as sets of individuals, binary relations of individuals, and, in general, k-ary relations of individuals.[1] Third-order languages are still more general, and so on, until the language of order ω, which is essentially the language of type theory.

The definition of the order of a structure was given previously, and we shall not repeat it here; intuitively, the greater the order of the structure, the greater the order of the elements that are relata of the involved relations (sets of sets, sets of sets of sets, and so on). Most scientific theories employ sophisticated mathematical theories, so they cannot be characterized as structures of *order-1*. A simple example is that of a topological space, which can be written as $\mathcal{T} = \langle X, \tau \rangle$, with X being a non-empty set and τ a *topology* over X, that is, a certain set of subsets of X obeying well-known postulates.[2] But, as we have seen already, more sophisticated structures, such as those for classical particle mechanics and non-relativistic quantum mechanics, to keep with our examples, require much more than simple relations whose relata are elements of the domains. So, in order to deal with interesting scientific theories, we need to employ higher-order structures.

In view of that need of higher-order structures, how is the semantic approach to theories to be framed? As we have seen, our understanding of the semantic approach is that, following the general lines of French

[Fre.15], it simply furnishes the tools for representing theories. The issue of whether there is the need to have an underlying language and how it affects the representational power of the approach is something we shall also touch on. The main point is that *language* is a problem when we *identify* a theory with some of its formulations. As we shall see, the approach by da Costa and Chuaqui as well as the approach by Suppes may deal with those issues. Thus we shall add some more philosophical remarks on both approaches and the relation between the postulates of their Suppes predicates with their respective models.

We begin with the approach by Suppes to theories.

6.2   FURTHER REMARKS ON THE SUPPES APPROACH

Let us recall that Suppes's approach to scientific theories involves specifying a wide class of structures that he calls the 'models' of a certain set-theoretical predicate. The predicate itself is understood as axiomatizing the theory so that its postulates are formulated in the language of set theory itself, perhaps extended with the symbols proper of the considered theory. Our examples in the previous chapter were classical particle mechanics, group theory, vector spaces theory, and non-relativistic quantum mechanics. For continuing the discussion and also for presenting a new example, let us consider the real vector space $\Re^n = \langle \mathbb{R}^n, \mathbb{R}, +, \cdot \rangle$, where $\mathbb{R}^n$ is the set of n-tuples of real numbers, and the operations are the addition of n-tuples (by summing the corresponding coordinates) and the product of n-tuples by real numbers. This structure is said to be a model of the theory of vector spaces. But what does this affirmation mean in precise terms? Are we really speaking about models in the same sense as Tarski?

The answer to that question will bring to light what is really distinctive about the approach by Suppes to the semantic approach. The main point is that, by lying the postulates of the theory directly in the language of set theory, and not in a specific formal language that is interpreted by the relevant structures, Suppes leaves behind precisely the semantic component, which is adopted when we have an interpretation of a language [Mul.11, sec.6], [KraAreMor.11], [Lut.15]. In the specific case of Suppes, there is simply no set Γ of postulates in a formal language so that the structures $\mathfrak{A}$ of a given class are such that they not only interpret the non-logical vocabulary of the language but also are such that

$$\mathfrak{A} \models \Gamma.$$

In fact, following Suppes's approach, that semantic component would be impossible, given that the postulates of Γ involve set-theoretical vocabulary and other elements. Recall that the postulates are formulated directly in the language of set theory. Any interpretation of that language would require an interpretation of the language of set theory itself (as we discussed in chapter 3, this can be done only if a stronger set theory is assumed as the metalanguage).

But then what is the relevant notion of model that is at stake? Certainly we have set-theoretical structures of any order, and the Suppes predicate serves to gather the relevant structures into classes that represent the theory. Given those classes, we can then study the relations between distinct theories, search for representation theorems, study the relation of the theory with data, and so on. But how do those structures 'model' the set-theoretical predicate if not in the Tarskian sense of satisfaction and truth?

The whole point is that we can prove *within* set theory that the structures in the relevant class have the properties demanded by the set-theoretical predicate. Let us consider, for instance, the case of the real vector space $\mathfrak{R}^n = \langle \mathbb{R}^n, \mathbb{R}, +, \cdot \rangle$. Following the traditional mathematical practice, we simply prove inside set theory itself that the structures 'satisfy' the postulates of vector space, which we represent by writing

$$\mathfrak{R}^n \models (\text{vector space axioms}),\tag{6.1}$$

where $\models$ is to be understood in the usual semantical consequence symbol. This means that, once we obtain the 'ingredients' of the structure in set theory, which are just sets (here we are avoiding the *Urelemente*), all we do is prove those sentences in the language of set theory as theorems of set theory (probably expanded by new symbols) that express the postulates of vector spaces for this particular sample. For instance, in order to show that $\mathfrak{R}^n \models \exists \mathbf{O} \forall \alpha (\mathbf{O} + \alpha = \alpha)$, we take $\mathbf{O} = \langle 0, \ldots, 0 \rangle$ and $\alpha = \langle x_1, \ldots, x_n \rangle$ and then show that

$$\mathbf{O} + \alpha = \langle 0, \ldots, 0 \rangle + \langle x_1, \ldots, x_n \rangle = \langle 0 + x_1, \ldots, 0 + x_n \rangle = \alpha.$$

That is, using the terminology with some care but in a very clear way, we show that

$$(\text{Set theory}) \vdash (\ \mathfrak{A} \models (\text{vector space axioms})\ ).$$

That is, as Suppes says, we are working with mathematics and not with metamathematics. This shows that the notion of *satisfaction* of the postulates by a structure, in Suppes's approach, is purely *syntactical*, consisting of mathematical proofs inside set theory. An even simpler example is the set ω of natural numbers, which is usually constructed inside set theory (ZFC, for instance). The typical proof that this set satisfies the Peano Postulates does not proceed in association with the relevant elements of the set with a formal language. Rather, what we usually do is provide for a proof inside set theory itself that there is a first element (the empty set), a sucessor function working as expected, and ω is inductive (see chapter 3 and [End.77] for further details).

Now, with these features in mind, we can characterize a scientific theory according to Suppes's 'semantic' view: let P be the set-theoretical predicate that axiomatizes a certain theory T, and $\mathfrak{A}$ a structure that satisfies the predicate. This structure is built with elements of a scale $\varepsilon(D)$, where D is the domain of the structure (in our sample case, $\mathbb{R}^n$). As a set (in most cases for scientific theories), each structure belongs to some V_α of the cumulative hierarchy, so, we can write in a short notation (suggested in [Mul.11]):

$$T = \{\mathfrak{A} \in V_\alpha : P(\mathfrak{A})\}, \tag{6.2}$$

which means that a theory is a collection (not a set of standard set theories, perhaps a proper class)[3] of structures that model the set-theoretical predicate.

Recall from the discussion in the previous chapter that a set-theoretical predicate, in the sense considered by Suppes, is able to select very specific classes of structures. It may provide for a selection of so-called intended models of a theory, leaving aside models that merely have the desired formal properties but that are of no relevance for the empirical study the theory was developed for. This will contrast with the classes of structures selected by the da Costa–Chuaqui approach, which we will explain soon, where one cannot 'exclude' some specific unintended structures from being inside the class comprising (representing) the theory.

So, in a sense, Suppes's own approach is not really a 'semantic' approach. Muller [Mul.11, sec.6] has noticed that, and Suppes [Sups.11] agreed. Then, even though Suppes himself mentioned the Tarskian tradition of models as encapsulating precisely the kind of model needed in science (for instance, in [Sups.62]), in the end, Suppes recognized that the models he needed are of a distinct nature. In fact, this approach is more likely to be useful when one is interested, as Suppes was, in the relation between theory and experiments. The simplicity of the approach leads one to deal more directly with such issues.

That last point is important for a discussion that was already raised in the first chapter: how can one make sense of the fact that we are 'axiomatizing a theory', an informal and pre-given theory T, and afterward, as it were, to develop a set-theoretical surrogate for it? Recall that there is a simple way out for that. In fact, following French [Fre.15], we are not claiming that an informal theory must be *identified* with its set-theoretical counterpart; the set-theoretical construction serves specific purposes, and in the case of Suppes, as he clearly stated in his [Sups.67], two of the main purposes are the search for a better understanding of the relation between data and theory and the study of theories even in the absence of standard axiomatizations. The main problem appears, recall, when one *reifies* a set-theoretical construction and then must face the battle of proving this to be the 'correct' reification against seemingly equivalent reifications.

Thus we have arrived at a possible characterization of scientific theories. This is the first step for the discussion of philosophical points of interest, such as the relation of theory with experience, the characterization of concepts such as empirical adequacy, and many others. Bas van Fraassen, for instance, advocates this approach and complements that we should look for *empirically adequate* theories [vanF.80], instead of the traditional search for 'true theories'. Here we will not enter this discussion; our aim is to proceed and present the da Costa–Chuaqui approach and, later, comment on how metamathematical issues impinge on such characterizations.

6.3   DA COSTA AND CHUAQUI, FURTHER REMARKS

Now, let us turn again to the approach by da Costa and Chuaqui to scientific theories. As we have already seen, differently from the approach of Suppes, this is a semantic approach in the traditional Tarskian style of semantics. However, differently from what is usually thought in the literature about the semantic approach, the relevant languages and structures involved need not be restricted to the elementary case, or *order-1* structures (recall our example of axiomatization of classical particle mechanics in the previous chapter). Higher-order languages and structures are allowed, and, in fact, they seem to be even required when it comes to most empirical theories.

What is more relevant is that, in this case, there is a sensible interpretation of a formal language (recall our discussion in the previous chapter about the language of a structure). It is in the language of the structure that one may write the postulates of the theory. In this sense, a theory is a class of models of a set of postulates written in a specific language. More precisely, given the language $\pounds$ for a species of structures, and given a set of postulates adequate $A_\pounds$ for that theory, the theory may be characterized by the class

$$\{\mathfrak{A} : \mathfrak{A} \models A_\pounds\}.$$

Here, recall, just as it happened in the case of the Suppes approach, there is no problem with the fact that we are using a theory, in the informal sense, to construct a set-theoretical representation for philosophical purposes. But once we have the approach by Suppes, what are the advantages of following this traditional approach with formal languages and axiomatization of the whole step theories?

This is an important question, and we have already mentioned some possible answers that it may suggest. Formalization of a theory brings with it much that may count as advantage. For instance, the discovery of unintended models (which is deemed a disadvantage by some adherents of the semantic approach) has proven very helpful in recent times (more on this soon, with relation to the metalanguage). For another advantage, recall that formalization opens the door for the use of formal methods that may allow us to obtain results that, at a purely informal level, are not easily achievable (see again chapter 2).

But perhaps the most interesting kind of result that the da Costa–Chuaqui approach allows concerns some problems related to theory equivalence. Recall from chapter 1 that a difficult challenge appears for those identifying a theory either with a class of models or else with a specific linguistic formulation: in the case of the semantic approach, alternative classes may reasonably claim the rights of representing the same theory; in the case of the syntactic approach, distinct linguistic formulations may be employed to characterize the same theory. Even those giving up identification, as we are doing, owe an account of how distinct representations may be reasonably employed to represent the same theory. As we discussed in chapter 1, the main general claim is that theory equivalence is achievable in the presence of a language for the structure. That is precisely what we have in the da Costa–Chuaqui approach.

The approach by da Costa and Chuaqui offers a great opportunity for the rigorous study of equivalent theories in the precise sense that we shall present now, rather briefly. Here we shall rely on the exposition on definability presented in the previous chapter and on the general development of those concepts by da Costa and Rodrigues [Cos.07]. In that work, da Costa and Rodrigues relate definability problems with certain kinds of invariance by automorphism in a structure. The group of automorphisms of a structure is the *Galois group* of that structure. The authors are able to generalize model theoretic notions and some tools of model theory for higher-order languages and structures. By considering infinitary higher-order languages as the language of our structure, following [Cos.07], we shall call a concept definable in this language as being *definable in the wide sense*. Now, two structures $\mathfrak{A}$ and $\mathfrak{A}'$ are *equivalent*, written $\mathfrak{A} \equiv \mathfrak{A}'$ if and only if the primitive relations of each of them are definable in terms of the primitive relations of the other. So we may think of two structures with distinct signatures as presenting the same theory only when they are equivalent in this sense.

What is more relevant for us now is that this notion of equivalence (which is somehow advanced in [Hal.12, Hal.15] and [Lut.15]) also has a model theoretic counterpart. As a consequence of the work by da Costa and Rodrigues, as well as da Costa and Bueno [Cos.11, p.157], we can state a theorem with necessary and sufficient conditions for equivalence: a given object r is definable in a certain vocabulary on the basis of a certain set of hypotheses Γ if and only if, for every structure $\mathfrak{A}$ modeling Γ, r is invariant by the Galois group of the restriction $\mathfrak{A}'$ of $\mathfrak{A}$ to the language without the symbol denoting r (for details, see [Cos.07, Cos.11]).

Now this may be a first step in the study of equivalence of theories. What is interesting is that the Galois group of structures (theories)[4] provides interesting techniques, relating syntactical aspects that are not always fully developed with model theoretic tools for the equivalence of theories. This is, as we mentioned, the first step in such studies, and may, perhaps, replace the debates on which approach is the 'correct' one: the syntactical or the semantic. They are both required, and the correct development of their relations may bring useful results for the philosophy of science as a discipline.

Obviously, this is a kind of metamathematical study of theories. This simply puts the metamathematical resources in the center of the stage in the philosophy of science again, or at least, in part of it. We shall now explore a little more about this aspect of the study of theories, while concentrating on the role of the metamathematical apparatus.

6.4   THE METAMATHEMATICAL FRAMEWORK

In order to illustrate why the consideration of the mathematics *where* the discussion is being developed is so relevant, let us advance some case studies as an attempt to raise the issue. This is what we shall do now.

It is well known that quantum mechanics needs unbounded operators, such as position and momentum operators [Hea.90]. An unbounded operator T is a linear operator (here supposed over a suitable Hilbert space $\mathcal{H}$) such that for any $M > 0$ there exists a vector α such that $\|T\alpha\| > M.\|\alpha\|$. Otherwise, T is bounded. Now consider the theory ZF+DC, where DC stands for a weakened form of the Axiom of Choice, entailing that a 'countable' form of the Axiom of Choice can be obtained. (In particular, if $\{B_n : n \in \omega\}$ is a countable collection of non-empty sets, then it follows from DC that there exists a choice function f with domain ω such that $f(n) \in B_n$ for each $n \in \omega$.) It can then be proven, as Solovay showed, that in ZF+DC (which is supposed to be consistent) the proposition "Every subset of $\mathbb{R}$ is Lebesgue measurable" cannot be disproved.[5] This proposition is false in standard ZFC. The same happens with the proposition "Each linear operator on a Hilbert space is bounded" [Mai.73]. This kind of result poses a difficulty to the friends of the semantic view: when we speak of the models of a scientific theory, such as quantum mechanics, which metamathematics should we use to define its models? Presumably, it cannot be Solovay's model in ZF+DC, since we need unbounded operators. So the choice of a suitable metamathematics is crucial. The characterization of a scientific theory according to the semantic approach is sensible to the mathematical apparatus employed in the very construction of those models!

As a second example, let us acknowledge that in the standard Hilbert space formalism for quantum mechanics, we deal with bases for the relevant Hilbert spaces. More specifically, we deal with orthonormal bases formed by eigenvectors of certain Hermitean operators. This is possible because we can prove, using the Axiom of Choice (which is part of the metatheory used here, essentially ZFU, so we are supposing the existence of *Urelemente*) that any Hilbert space $\mathcal{H}$ has a basis. Moreover, it can also be shown that each basis has a specific cardinality, which is the same for all bases of $\mathcal{H}$ (this is defined to be the dimension of the space). But in certain set theories in which the Axiom of Choice does not hold in full generality, such as in Laüchi's permutation models, we obtain: (a) vector spaces with no basis, and (b) a vector space that has two bases of different cardinalities [Jec.77, p.366]. Now if a vector space has no basis, it seems that it cannot be used as part of the standard formalism of quantum mechanics. The latter formalism presupposes the availability of suitable bases. As a result, again we see it very clearly that the theory, understood as a class of models, depends crucially on the metamathematics that is used. Of course it seems interesting for the philosophy of science to study such models as well as possible models of a scientific theory.

Our third example also comes from quantum mechanics (for details, see [KraBue.07]). In order to keep the example self-contained, we recall that in the 1920s, Thoralf Skolem noticed that there are concepts that are, as it were, the same in all models of, say, ZFC (these concepts are termed *absolute*). For instance, the concept of 'ordinal' does not change from model to model as the concept of 'cardinal' does (these are called *relative* concepts. For example, if ZFC is formulated as a first-order theory, if consistent, it has a countable model due to the Lowënheim-Skolem theorem. But, in this model, the set of real numbers, which can be constructed in ZFC (say by Dedekind cuts), must be countable—a fact that seems to contradict 'Cantor's theorem', according to which there is no bijection between the set of real numbers and the set of natural numbers. However, as Skolem himself noticed, this result does not lead to a 'real' paradox, for the bijection may exist *outside* the countable model, and not as a set of ZFC proper [Sko.22]. But, in the *standard* model of ZFC, the cardinality of the set of the real numbers cannot be denumerable, as Cantor's theorem teaches us. This result shows that the set of real numbers may have different cardinalities depending on the model we consider, and this happens in general for other sets.

The relativism of the notion of a cardinal number can be seen with this simple example.[6] Suppose that our theory makes use of the set $A = \{2^{\aleph_0}, \aleph_\alpha\}$, where α is a cardinal. How many elements does this set have? Well, it depends on the model of set theory we are considering. Really, if the Continuum Hypothesis holds,[7] the set has just one element, but it has more than one otherwise (say, when $2^{\aleph_0} = \aleph_\omega$).

Let us take this result in mind for what follows. After presenting his modal interpretation of quantum mechanics, Bas van Fraassen addresses the problem of identical particles in quantum physics, which he regards as one of the three main issues in the philosophical discussion on quantum mechanics [vanF.91, p.133]. And he notes, "identical particles [...] are certainly qualitatively the same, in all the respects represented in quantum-mechanical models—yet still numerically distinct" (ibid., p. 376). In a previous paper, he was still more explicit, insisting that "if two particles are of the same kind, and have the same state of motion, nothing in the quantum-mechanical description distinguishes them. Yet this is possible" [vanF.84].

Van Fraassen's quotations are intriguing. Particles of the same kind and in the same state of motion are 'identical', in physicists' jargon, and according to their standards, nothing can distinguish them. So if they cannot be distinguished *within* the quantum-mechanical formalism, how can they still be distinguished at all? The answer, we suggest—following the parallel case made by Skolem in set theory—is that the particles can be distinguished *outside* the framework of quantum mechanics. But what does this mean? As we have seen, in the foundations of set theory, considerations regarding what holds inside or outside a certain model are quite common. But can we make sense of this way of speaking in the philosophy of science as well?

Van Fraassen's modal interpretation of quantum mechanics takes quantum propositions as modal statements, which give "first and foremost about what can and what must happen, and only indirectly about what actually does happen" [vanF.80]. In other words, the modal account, by offering an interpretation of quantum mechanics, spells out how the world could be if quantum mechanics *were* true [vanF.91, p.242]. To motivate his proposal, van Fraassen recalls one of the most intriguing features of quantum physics, namely, the sense in which quantum mechanics is an indeterministic theory. Although the dynamics of an isolated system evolves according to Schrödinger's equation (hence deterministically), the system as a whole cannot be analyzed in terms of its component parts. So, apparently, the quantum-mechanical state of the whole system contains only incomplete information about the system. Bohr's proposal, recalls van Fraassen, emphasizes that it is still possible to have complete information about the system, given that the states of the system's components and the state of the whole system do not determine each other [vanF.80]. As a result, on the basis of the state of a complete system $X + Y$, we can in general ascribe at most mixed states to X and Y, but from them nothing can be said back about the state of the whole system. As van Fraassen notes:

> [I]f we can predict the future states of an isolated system on the basis of its present state [by means of Schrödinger's equation], then how can we be ignorant about the future events involving its components unless the information in those total states is incomplete? For surely any true description of a component is a partial but true description of the whole?
> (*ibid.*)

Van Fraassen's answer is obtained from a distinction between quantum dynamical states and experimental events. The former are what a vector or a statistical operator represents. They are things completely *embedded* in the theory, whose evolution is governed by dynamical laws. In other words, we can say that dynamical states are described *within* the formalism of quantum mechanics. Events, on the contrary, are *extra-theoretic* entities that satisfy the probability calculations in the sense that physicists do in quantum theory.

The same conceptual distinction can be drawn by distinguishing between *state attributions* and *value attributions* of a physical system. The former is a theoretic construct, and part of the challenge involved in theory's construction depends upon a proper representation of these states. Value attributions, in turn, express values that an observable actually has. Since the point is important for our argument, let us consider it in more detail.

A value state is specified by stating which observables have values and what they are. A value-attributing proposition then states that an observable m actually has a value in a (Borel) set E (in symbols, $\langle m, E \rangle$. The connection between them is that value states are truth-makers of value attributing propositions [vanF.91, pp.275–6]. On the other hand, we have the dynamic state, which states how the system will evolve, either in isolation or in interaction with another system. A state-attributing proposition then states that a measurement of an observable m must have a value in a (Borel) set E (in symbols, $[m, E]$. Again, dynamic states and state-attributing propositions are connected by the fact that the former are what make the latter true. Now the crucial feature of the modal account is to distinguish value- and state-attributing propositions. The motivation for this distinction comes from difficulties faced by the standard interpretation of quantum mechanics, as articulated by von Neumann, for not distinguishing them.

Von Neumann's interpretation of quantum mechanics identifies these two concepts. After all, not only does von Neumann consider that a system can be said to posses a value of a certain variable when it is in an eigenstate of the corresponding observable, but he also accepts that if the state vector is not an eigenstate of some observable, then it has no value at all (see also [Bit.96, p.149]). In this case, the system is supposed to be characterized by a well-defined value of the observable when the probability is 1. But if this probability is not 1, then the observable is supposed to have no value at all. To remove this discontinuity, van Fraassen offers an account according to which probability ascriptions are not equivalent to value ascriptions (*ibid.*).

On von Neumann's interpretation, attributions of values and classification of states are closely related: an observable B has value b if and only if a B-measurement is certain to have outcome b (where b is a real number). The problem here is that in order to accommodate states for which measurement has uncertain outcomes, von Neumann made a radical move: if the outcome of a measurement of B is uncertain, B has no value at all [vanF.91, p.274]. To avoid this answer, the modal interpretation introduces the distinction between values and states. With this distinction in place, the introduction of 'unsharp' values of observables is allowed. And this is how the possibility of uncertain outcomes in measurement can be accommodated.[8] As van Fraassen points out [vanF.91, pp. 280–1], if a physical system X has dynamic state (represented by an operator) W at a time t, the state-attributions $[M, E]$ that are true are those such that $\text{Tr}(WI_E^M) = 1$.[9] As opposed to state-attributions, value-attributions cannot be deduced from the dynamic state. But, according to van Fraassen, they are constrained in three ways:[10] (i) if $[M, E]$ is true, so is the value-attribution $\langle m, E \rangle$; that is, observable M has value in E; (ii) all true value-attributions could have probability 1 together; and (iii) the set of true value-attributions is maximal with respect to feature (ii) [vanF.91, p. 281]. So the assignment of truth-conditions to state- and value-attributing propositions is crucial to spell out the difference between them (the former, but not the latter, can be deduced from the dynamic state).

To sum up, there is an important distinction between state-attribution and value-attribution, or between states and events, and this distinction cannot be reduced to something more basic. States, as already noted, are described *in* the scope of (the formalism of) quantum mechanics by vectors of an appropriate Hilbert space, while events are not. After all, events are statements such as *Observable B pertaining to system X has value b*, and such events are described if they are assigned probabilities, but "they are not the same thing as the states which assign them probabilities" [vanF.91, p. 279].

The distinction between states and events is similar to the distinction between absolute and relative notions in set theory discussed earlier, at least in the following way: we are contrasting *intra-theoretic* properties with properties that hold *outside* the models under consideration. It's curious that when Skolem introduced his 'paradox', he intended to use it to show the inadequacy of set theory as a foundation for mathematics. The outcome, however, was precisely the opposite. His result was incorporated as part of the rich conceptual framework offered by set-theoretic notions. Similarly, van Fraassen developed the modal interpretation of quantum mechanics as part of a defense of an empiricist view. In the end, however, the modal interpretation became part the revival of realist interpretations of quantum theory.

It should now be clear that both in the philosophy of science and in the foundations of set theory there is room for discussing what holds 'inside' a particular model (or formalism) and what holds 'outside' the model (formalism). If a theory is presented as a class of models, it makes perfect sense to ask whether there are concepts that remain the same in all models and concepts that change from model to model—that is, that have a certain extension 'inside' a model, but a different one in another model or when considered outside the model.

Finally, let's examine another case. Consider the concept of indistinguishable (or indiscernible) objects. The idea of indiscernibility is of fundamental importance in contemporary physics (for a historical account and further discussion, see [FreKra.06]). Standard mathematics and classical logic imply that every object is an individual, in the sense that each object can always be distinguished from any other at least in principle, and these individuals *retain* their individuality even when mixed with others of similar species.[11] As a result, to accommodate indistinguishable objects, some mathematical trick needs to be introduced.

In quantum mechanics, this is done by imposing some kind of symmetry condition. Suppose we are to describe how two identical bosons, 1 and 2, can be distributed in two possible states, A and B. As is well known, the vectors in the relevant Hilbert space are $|\psi_1^A\rangle|\psi_2^A\rangle$, which states that both bosons are in A, while $|\psi_1^B\rangle|\psi_2^B\rangle$ states that both are in B, and $\frac{1}{\sqrt{2}}|\psi_1^A\rangle|\psi_2^B\rangle + \frac{1}{\sqrt{2}}|\psi_2^A\rangle|\psi_1^B\rangle$ states that *one* of them is in A and *the other* is in B. Thus the indistinguishability between bosons 1 and 2 (in the third case) emerges from the symmetry of the function, which is invariant by permutations of the labels. Some people claim that the individuality of quantum objects is then lost. According to our point of view, there is nothing to be lost, for in one of the possible approaches to the subject, these objects do not have identity conditions to start with [FreKra.06]. The artificiality of the problem is that these objects were first assumed to be individuals by their labels 1 and 2. Thus, by an adequate choice of the relevant vectors, *we have made them* indiscernible. However, the objects cannot be said to be indiscernible *outside* the framework (say, ZFC), since *we can* distinguish them—e.g., by their labels 1 and 2. In this way, the notion of indistinguishable object seems to be *relative*.

The mathematical trick we used consists of limiting the discourse to the scope of a certain set-theoretic structure (as we saw, the models of quantum physics can be taken to be structures of this kind). We then consider as indiscernible those objects that are invariant by the automorphisms of the structure. Now in ZFC any structure can be extended to a rigid structure, that is, to a structure where the only automorphism is the identity function. Hence, in the rigid structure (the *whole* ZFC model $\mathcal{V} = \langle V, \in \rangle$ is rigid), any object is an individual. In short, there are no truly indiscernible objects in standard mathematics (and logic). A suitable mathematics for developing quantum mechanics taking the non-individuality of quantum objects *from the start* needs to be taken into account; we have developed *quasi-set theory* for that, but this subject extrapolates the contents of this book (the interested reader can have a look at [FreKra.06]; [KraAre.15]).

NOTES

1. As before, here we concentrate on relations only, since distinguished elements and operations can be viewed as particular relations.
2. The postulates are: (1) $\emptyset$ and X belong to τ, (2) any finite intersection of elements of τ belongs to τ, and (3) any reunion of elements of τ belongs to τ, which can be seen in any book of topology (the first postulate is implied by the other two [Bou.95, chap.1]).

3. In the von Neumann-Bernays-Gödel (NBG) set theory, the primitive elements are *classes*, and *sets*, which are copies of the ZFC sets, are classes that belong to other classes. The classes that do not belong to other classes are called *proper classes*. These entities lie outside ZFC.

4. The Galois group of a theory is the Galois group of the structures that model the set-theoretical predicate that axiomatize the theory.

5. The Lebesgue measure generalizes the usual notion of 'measure' of a set, say the length of an interval, the area of a plane figure, the volume of a tridimensional figure. The precise definition is not necessary here.

6. We owe this example to A.M.N. Coelho.

7. Cantor's Continuum Hypothesis says that there is no set of real numbers with cardinality between the cardinality of the natural numbers ($\aleph_0$) and the cardinality of the set of real numbers, $\mathscr{C}$ (for 'continuum'). But Cantor himself proved that $2^{\aleph_0} = \mathscr{C}$, and postulated that the *next* cardinal number after $\aleph_0$, termed $\aleph_1$, is precisely $\mathscr{C}$. The *generalized* continuum hypothesis (GCH) says that $2^{\aleph_\alpha} = \aleph_{\alpha+1}$; it was known from Sierpinski that ZF+GCH implies the Axiom of Choice. It was later shown by Gödel and Cohen that this hypothesis is independent of the axioms of ZF (without choice).

8. Note that the value-state distinction is cashed out in terms of the concept of truth. The relationship between these ideas and the concept of quasi-truth is developed in [Bue.00].

9. A few comments about the notation: (a) Tr is a linear functional of operators into numbers (the trace map), which gives us the probability that a measurement of the observable m has a value in the Borel set E; (b) I_E^M is an Hermitean operator such that $I_E^M(x) = x$ if $M(x) = ax$ for some $a \in E$ and is the null vector if $M(x) = bx$ for some value $b \notin E$, where M is the Hermitian operator, which represents m; and (c) that the trace function Tr provides a probability is due to the fact that $P_x^m(E) = (x \cdot I_E^M x) = \text{Tr}(I_x I_E^M)$, where $P_x^m(E)$ is the probability that a measurement of m has a value in E, $(x \cdot I_E^M x)$ is the inner product of x and $I_E^M x$ and I_x is the projection on the subspace $[x]$ spanned by x. For details, see [vanF.91, pp. 147–52, 157–65, and 280–1].

10. Which again are spelled out in terms of truth.

11. Although the concepts of individuality and distinguishability should not be confused; see [FreKra.06].

Bibliography

[All.38] Allen, E. S. 1938, *Review of The Axiomatic Method in Biology*, by J. H. Woodger. Cambridge, University Press, 1937. 10+174 p. *Bull. Amer. Math. Soc.*, Vol. 44 (11): 763

[AreKra.14] Arenhart, J.R.B. and Krause, D. 2014, From primitive identity to the non-individuality of quantum objects. *Studies in History and Philosophy of Modern Physics* 46 (B): 273

[Aris.89] Aristotle 1989, *Prior Analytics*. Trans. Robin Smith. Cambridge: Indianapolis, Hackett Pu.

[Arn.97] Arnol'd W., 1995, Will mathematics survive? *The Mathematical Intelligencer* 7 (3), 6

[BarMac.75] Barnes, D. W. and Mack, J. M. 1975, *An Algebraic Introduction to Mathematical Logic*. New York, Heidelberg, Berlim: Springer-Verlag.

[Bell.09] Bell, J. L. 2009, 'Infinitary Logic', *The Stanford Encyclopedia of Philosophy* (Spring 2009 Edition), URL = <http://plato.stanford.edu/archives/spr2016/entries/logic-infinitary/>. Access date: 10/04/2016.

[Bit.96] Bitbol, M. 1996. *Schrödinger's philosophy of quantum mechanics*. Dordrecht: Kluwer Academic Publishers.

[Bla.14] Blanchette, P. 2014, The Frege-Hilbert Controversy, *The Stanford Encyclopedia of Philosophy* (Spring 2014 Edition), Edward N. Zalta (ed.), URL = <http://plato.stanford.edu/archives/spr2014/entries/frege-hilbert/>. Access date: 01/08/2015.

[Bli.88] Blizard, W. D. 1988, Multiset theory. *Notre Dame Journal of Formal Logic* 30 (1): 36–66.

[Bog.79] Bogdan, R. 1979, *Patrick Suppes*. Dordrecht: D.Reidel.

[Bou.50] Bourbaki, N. 1950, The architecture of mathematics. *The American Mathematical Monthly*, Vol. 57, No. 4, p. 221

[Bou.68] Bourbaki, N. 1968, *Theory of Sets*. Paris: Hermann and Addison-Wesley.

[Bou.90] Bourbaki, N. 1990, *Théorie des Ensembles*. Paris: Masson.

[BrigCos.71] Brignole, D. and da Costa, N.C.A. 1971, On Supernormal Ehresmann-Dedecker Universes. *Mathematische Zeitschrift* (122), p. 342

[Bue.00] Bueno, O. 2000, Quasi-truth in quasi-set theory. *Synthese* 125: 33–53.

[Can.55] Cantor. G. 1955, *Contribution to the Founding of the Theory of Transfinite Numbers*. New York: Dover.

[Cao.99] Cao, T. Y. 1999, *Conceptual Foundations of Quantum Field Theory*. Cambridge: Cambridge Un. Press.

[Car.58] Carnap, R. 1958, *Introduction to Symbolic Logic and Its Applications*. New York: Dover.

[Con.06] Contessa, G. 2006, Scientific models, partial structures and the new received view of theories. *Studies in History and Philosophy of Science* 37: 370–77.

[Cor.92] Corry, L. 1992, Nicolas Bourbaki and the concept of mathematical structure, *Synthese* 92 (3): 315–48.

[Cor.04] Corry, L. 2004, *Modern Algebra and the Rise of Mathematical Structures.* 2nd.ed., Basel: Springer.

[Cos.80] da Costa. N.C.A. 1980, *Ensaio sobre os fundamentos da lógica* (in portuguese). 1st ed. São Paulo: Hucitec.

[Cos.07] da Costa, N.C.A. 2007, *Abstract Logics*, Manuscript, Federal University of Santa Catarina (not published).

[Cos.11] da Costa, N.C.A. and Bueno, O. 2011. Remarks on Abstract Galois Theory. *Manuscrito* 341: 151

[CosChu.88] da Costa, N.C.A. and Chuaqui, R. 1988, 'On Suppes' set theoretical predicates', *Erkenntnis* 29, 95

[CosFre.03] da Costa, N.C.A. and French, S. 2003, *Science and Partial Truth: A Unitary Approach to Models and Scientific Reasoning.* Oxford: Oxford Un. Press.

[CosRod.07] da Costa, N.C.A. and Rodrigues, A.M.N. 2007, 'Definability and invariance', *Studia Logica* 82, 1–30.

[CosDor.08] da Costa, N.C.A. and Doria, F. A. 2008, *On the Foundations of Science*, COPPE/UFRJ, Cadernos do Grupo de Altos Estudos, Vol. II. Rio de Janeiro: Editora E-Papers.

[CosKraBue.06] da Costa, N.C.A., Krause, D. and Bueno, O. 2006, 'Paraconsistent logics and paraconsistency', in in D. Jacquette, D. M.Gabbay, P.Thagard and J. Woods (eds.), *Philosophy of Logic*, Amsterdam: Elsevier, in the series Handbook of the Philosophy of Science, v. 5, 655–781.

[CovHaw.04] Cover, J. A. and O'leary-Hawthorne, J. 2004, *Substance and Individuation in Leibniz.* Cambridge, Cambridge Un. Press.

[DalGiuGre.04] Dalla Chiara, Giuntini, R., and Greechie, R. 2004, *Reasoning in Quantum Theory: Sharp and Unsharp Quantum Logics.* Dordrecht: Kluwer Ac. Press (Trends in Logic, vol.22).

[DalTor.93] Dalla Chiara, and Toraldo di Francia, G. 1993, 'Individuals, kinds and names in physics', in Corsi, G., Dalla Chiara, M. L. and Ghirardi, G. C. (eds.), *Bridging the Gap: Philosophy, Mathematics, and Physics,* Dordrecht: Kluwer Ac. Press (Boston Studies in the Philosophy of Science, 140), 261

[Dan.et al.02] Daniel F. Styer, Miranda S. Balkin, Kathryn M. Becker, Matthew R. Burns, Christopher E. Dudley, Scott T. Forth, Jeremy S. Gaumer, Mark A. Kramer, David C. Oertel, Leonard H. Park, Marie T. Rinkoski, Clait T. Smith, Timothy D. Wotherspoon. 2002. Nine formulations of quantum mechanics. *Am. J. Physics* 70 (3): 288–97.

[Dau.90] Dauben, J. W. 1990, *Georg Cantor: His Mathematics and Philosophy of the Infinite.* Princeton: Princeton Un. Press.

[D'Es.03] D'Espagnat, B. 2003, *Veiled Reality: An Analysis of Present-Day Quantum Mechanical Concepts.* Boulder, CO: Westview Press.

[D'Es.06] D'Espagnat, B. 2006, *On Physics and Philosophy.* Princeton: Princeton University Press.

[DomHol.07] Domenech, G. and Holik, F. 2007, 'A discussion on particle number and quantum indistinguishability', *Foundations of Physics* 37 (6): 855

[DomHolKra.08] Domenech, G., Holik, F. and Krause, D. 2008, 'Q-spaces and the foundations of quantum mechanics', *Foundations of Physics* 38 (11) 969.
[DomHolKniKra.08]

[Dou.11] Douven, I. 2011, 'Abduction', *The Stanford Encyclopedia of Philosophy* (Spring 2011 Edition), Edward N. Zalta (ed.), URL = <http://plato.stanford.edu/archives/spr2011/entries/abduction/>. Access date: 01/08/2015.

[DutRec.15] Dutilh Novaes, C. and Reck, E. 2015, Carnapian explications, formalisms as cognitive tools, and the paradox of adequate formalization. Forthcoming in *Synthese*, DOI 10.1007/s11229–015–0816-z, 2015.

[EinInf.38] Einstein, A. and Infeld. L. 1938, *The Evolution of Physics*. New York: Simon and Schuster.

[End.77] Enderton, H. B. 1977, *Elements of Set Theory*. New York: Academic Press.

[Esan.13] Esanu, A. 2013, Evolutionary biology and the axiomatic method revisited, *The Romanian Journal of Analytic Philosophy*, Vol. VII (1): 19–41.

[Eucl.08] Euclid 2008, *Euclid's Element of Geometry*. English translation, by Richard Fitzpatrick http://farside.ph.utexas.edu/Books/Euclid/Elements.pdf. Access date: 01/08/2015.

[Fra.66] Fraenkel, A. A. 1966, *Abstract Set Theory* Amsterdam: North Holland.

[FraBarLev.73] Fraenkel, A. A., Bar-Hillel, Y. and Levy, A. 1973, *Foundations of Set Theory*. 2nd. ed. Amsterdam and London: North-Holland.

[Fra.82] Franco de Oliveira, A. J. 1982, *Teoria de Conjuntos, Intuitiva e Axiomática* Lisboa: Livraria Escolar Editora.

[Fre.15] French, S. 2015, (Structural) realism and its representational vehicles. *Synthese*, DOI: 10.1007/s11229–015–0879-x, 2015.

[FreKra.06] French, S. and Krause, D. 2006, *Identity in Physics: A Historical, Philosophical, and Formal Analysis* Oxford: Oxford Un. Press.

[FreKra.09] French, S. and Krause, D. 2009, Remarks on the theory of quasi-sets. *Studia Logica* 95 (1–2): 101–12.

[FreLad.03] French, S. and Ladyman, J. 2003, Remodelling structural realism: quantum physics and the metaphysics of structure *Synthese* 36 31–56.

[GlaGio.00] Glashow, S. L. Georgi, H. 2000, La teoría unificada de las fuerzas de las partículas elementales. In Glashow, S. L., *El Encanto de la Física*. Barcelona: Tusquets. Original *The Charm of Physics*, 1991.

[Ger.85] Geroch, R. 1985, *Mathematical Physics* Chicago: Chicago Un. Press.

[Gra.08] Gratzer, G. A. 2008, *Universal Algebra* New York: Springer.

[Gra.00] Gray, J. J. 2000, *The Hilbert Challenge* Oxford: Oxford Un. Press.

[Gro.99] Gross, D., 1999, The triumph and limitations of quantum field theory. In [Cao.99], pp. 56–67.

[Hal.12] Halvorson, H. 2012, What scientific theories could not be. *Philosophy of science* 79 (2) 183–206.

[Hal.15] Halvorson, H. 2015, Scientific Theories. Forthcoming in *The Oxford Handbook of Philosophy of Science*.

[Hea.90] Heathcote, A. 1990, Unbounded operators and the incompleteness of quantum mechanics. *Philosophy of Science* 57: 523–34.

[Hei.89] Heisenberg, W. 1989, *Physics and Philosophy*. London: Penguin Books.

[Hen.et al.59] Henkin, L., Suppes, P. & Tarski, A. (eds.) 1959, *The Axiomatic Method with Special Reference to Geometry and Physics*. Amsterdam: North-Holland.

[Hil.30] Hilbert, D. 1930, 'Address to the Society of German Scientists and Physicians', 8th September 1930.

[Hil.50] Hilbert, D. 1950, *Foundations of Geometry* (English translation by E. J. Townsend), La Salle, Open Court.

[Hil.76] Hilbert, D. 1976, Mathematical Problems. In Browder, F. E. (ed.), *Proceedings of Symposium in Pure and Applied Mathematics: Mathematical Problems Arising from Hilbert Problems*. Providence: American Mathematical Society, Vol.CCVIII, Part. I, pp. 1–34.

[Hil.96] Hilbert, D. 1996, Axiomatic thought. In Ewald, W. B. (ed.), *From Kant to Hilbert. A Source Book in the Foundations of Mathematics, vol. 2*. Oxford: Oxford University Press, pp. 1089–1096.

[HilAck.50] Hilbert, D. and Ackermann, W. 1950, *Principles of Mathematical Logic.* Providence: American Mathematical Society.

[Hod.01] Hodges, W. 2001, Elementary Predicate Logic. In Gabbay, Dov. M., and Guenthner, F., (eds.) *Handbook of Philosophical Logic*, vol. 1, 2nd. ed., 2001, Dordrecht: Springer.

[Hof.79] Hoffstadter, D. R. 1979, *Gödel, Escher, and Bach: an Eternal Golden Braid.* New York: Basic Books.

[Ign.96] Ignatieff, Y. (ed.) 1996, *The Mathematical World of Walter Noll.* Berlin and Heidelberg: Springer-Verlag.

[Jam.62] Jammer, M. 1962, *The Concepts of Force.* New York: Harper Torchbook.

[Jam.74] Jammer, M. 1974, *The Philosophy of Quantum Mechanics.* New York: John Wiley & Sons.

[Jau.68] Jauch, J. M 1968, *Foundations of Quantum Mechanics.* Cambridge: Addison Wesley.

[Jec.77] Jech, T. 1977, About the axiom of choice. In J. Barwise (ed.), *Handbook of Mathematical Logic* Amsterdam: North-Holland, pp. 345–70.

[JonBet.10] Jong, R. de and Betti, A. 2010, The classical model of science: a millennia-old model of scientific rationality. *Synthese* 174: 185–203.

[Jong.85] Jongeling, T. B. 1985, On an axiomatization of evolutionary theory, *J. Theoretical Biology* 117: 529–43.

[Kne.63] Kneebone, G. T. 1963, *Mathematical Logic and the Foundations of Mathematics.* Dordrecht: Van Nostrand.

[KraAre.15] Krause, D. and Arenhart, J.R.B. 2015, Individuality, quantum physics, and a metaphysics of non- individuals: the role of the formal. In A. Guay and T. Pradeu (eds.) Individuals Across the Sciences. Oxford: Oxford Un. Press, pp. 61–70.

[KraAreMor.11] Krause, D., Arenhart, J.R.B. and Moraes, F.T.F. 2011, Axiomatization and models of scientific theories. *Foundations of Science*, 16: 363–82.

[KraBue.07] Krause, D. and Bueno, O. 2007. Scientific theories, models, and the semantic approach. *Principia* 11 (2): 187–201

[Kun.09] Kunen, K. 2009, *The Foundations of Mathematics*, London: College Pu.

[Lad.57] Ladrière, J. 1957, *Le Limitations Internes des Formalismes.* Louvain: Nauwelaerts; Paris: Gauthier-Villars.

[Lak.76] Lakatos, I. 1976, *Proofs and Refutations: The Logic of Mathematical Discovery.* Cambridge: Cambridge Un. Press.

[Lan.66] Landau, E. 1966, *Foundations of Analysis.* New York: Chelsea Pu. Co.

[Lei.95] Leibniz. G. W. 1995, *Philosophical Writings.* London and Rutland: Everyman.

[Low.14]Low, Z. I. 2014, Universes for category theory. http://arxiv.org/pdf/1304.5227.pdf. Access date: 10/08/2015.

[Lut.12] Lutz, S. 2012, On a Straw Man in the Philosophy of Science: a Defense of the Received View. *HOPOS: The Journal of the International Society for the History of Philosophy of Science* 2 (1): 77–120.

[Lut.15] Lutz, S. 2015, What Was the Syntax-Semantics Debate in the Philosophy of Science About? *Philosophy and Phenomenological research*, doi: 10.1111/phpr.12221.

[Lui.97] Lui, S. H. 1997, An interview with Vladimir Arnol'd. *Notices of the AMS* 44 (4): 432–8 (April).

[Mac.63] Mackey, G. W. 1963, *Mathematical Foundations of Quantum Mechanics.* Reading, MA: Benjamin, Advanced Book Program.

[Mac.98] Mac Lane, S. 1998, *Categories for the Working Mathematician* (Graduate Texts in Mathematics). New York: Springer.

[Mad.05] de la Madrid, R. 2005, The role of the rigged space in quantum mechanics. In arXiv:quant-ph/0502053

[MagKra.01] MagalhÃes J. C. and Krause D. 2001, Suppes predicate for genetics and natural selection, *J Theor Biol.* 21, 209 (2): 141–53.

[Mai.73] Maitland Wright, J. D. 1973, All operators on a Hilbert space are bounded, *Bulletin of the Americal Mathematical Society* 79 (6): 1247–50.

[Man.10] Manin, Yu. I. 2010, *A Course in Mathematical Logic for Mathematicians*, 2nd. ed. New York: Springer

[Mar.93] Marchisotto, E. 1993, Mario Pieri and his contributions to geometry and foundations of mathematics, *Historia Mathematica* 20: 285–303.

[McSS.53] McKinsey, J.C.C., Sugar, A. C., Suppes, P. 1953, Axiomatic foundations of classical particle mechanics. *Journal of Rational Mechanics and Analysis*, Vol.2, No.2 (April): 253–72.

[Mar.07] Marquis, J. -P. 2007, Category theory, *Stanford Enc. Philosophy*, http://plato.stanford.edu/entries/category-theory. Access date: 01/08/2015.

[Men.97] Mendelson, E. 1997, *Introduction to Mathematical Logic.* Cornwall: Chapman & Hall.

[Moo.82] Moore, G. H. 1982, *Zermelo's Axiom of Choice: Its Origins, Developments, and Influence.* New York, Heidelberg and Berlin: Springer-Verlag.

[Mul.11] Muller, F. A. 2011, Reflections on the revolution at Stanford. *Synthese*, 183 (1): 87–114.

[MulSau.08] Muller, F. A. and Saunders, S. W. 2008, Discerning fermions, *British Journal for the Philosophy of Science* 59: 499–548.

[MulSee.09] Muller F. A., Seevinck M. P. 2009. Discerning elementary particles. *Philosophy of Science* 76, 179–200.

[Nac.07] Nachotomy, O. 2007, *Possibility, Agency, and Individuality in Leibniz's Metaphysics.* New York: Springer.

[Pen.89] Penrose, R. 1989, *The emperor's new mind.* Oxford: Oxford Un. Press.

[Pen.05] Penrose, R. 2005, *The Road to Reality: A Complete Guide to the Laws of the Universe.* New York: Alfred A. Knoff.

[Pop.72] Popper, K. R. 1972, *Objective Knowledge: An Evolutionary Approach.* Oxford: Clarendon Press.

[Prz.69] Przelecki, M. 1969, *The Logic of Empirical Theories.* London: Routledge & Kegan Paul (Monographs in Modern Logic Series).

[Pug.87] Pugorelov, A. 1987, *Geometry.* Moscow: Mir.

[Red.87] Redhead, M. 1987, *Incompleteness, Nonlocality and Realism: A Prolegomenon to the Philosophy of Quantum Mechanics.* Oxford: Clarendon Press.

[RubRub.70] Rubin, H and Rubin, J. E. 1970, *Equivalents of the Axiom of Choice.* Amsterdam and London: North-Holland, 2nd. printing.

[Sau.98] Sauer, T. 1998, The relativity of discovery: Hilbert's first note on the foundations of physics. In arXiv:physics/9811050v1

[Sch.95] Schrödinger, E. 1995, *The Interpretation of Quantum Mechanics: Dublin Seminars(1949–1955) and Other Unpublished Essays.* Woodbridge: Ox Bow Press.

[Sha.91] Shapiro, S., 1991, *Foundations without Foundationalism, a Case for Second-Order Logic.* Oxford: Clarendon Press.

[Sho.67] Shoenfield, J. R. 1967, *Mathematical Logic.* Reading: Addison Wesley. (reprinted by the Association of Symbolic Logic, 2000).

[Sho.77] Shoenfield, J. R. 1977, Axioms of set theory. In J. Barwise (ed.) 1977. *Handbook of Mathematical Logic.* Amsterdam: North-Holland, pp. 321–44.

[Sim.87] Simons. P. 1987, *Parts: A Study in Ontology.* Oxford: Oxford University Press.

[Sko.22] Skolem, T. 1922. Some remarks on axiomatized set theory. Reproduced in J. van Heijenoort (ed.) 1967. *From Frege to Gödel.* Cambridge: Harvard University Press: 290–301.

132 Bibliography

[Smu.91] Smullyan, R. 1991, *Gödel's Incompleteness Theorems*. Oxford: Oxford Univ. Press.

[Steg.79] StegmÜller, W. 1979, *The Structuralist View of Theories. A Possible Analogue of Bourbaki Programme in Physical Sciences*. Berlin and Heidelberg: Springer-Verlag.

[Sup.77] Suppe, F. (ed.) 1977, *The Structure of Scientific Theories*. Urbana: Chicago Un. Press.

[Sup.00] Suppe, F., 2000, Understanding scientific theories: an assesment of developments, 1969–1998. *Philosophy of science* 67, Supplement, pp. S102-S115.

[Sups.57] Suppes, P. 1957, *Introduction to Logic*. Princeton: Van Nostrand.

[Sups.60] Suppes, P. 1960, A Comparison of the Meaning and Uses of Models in Mathematics and the Natural Sciences, *Synthese* 12: 287–301.

[Sups.62] Suppes, P. 1962, Models of Data. In E. Nagel, P. Suppes, and A. Tarski (eds.), *Logic, Methodology, and Philosophy of Science: Proceedings of the 1960 International Congress*. Stanford: Stanford Un. Press, pp. 252–61.

[Sups.67] Suppes, P. 1967, What is a scientific theory? In Morgenbesser, S. (ed.), *Philosophy of Science Today* New York: Basic Books, pp. 55–67.

[Sups.77] Suppes, P. 1977, The structure of theories and the analysis of data. In: Suppe, F. (ed.), *The Structure of Scientific Theories*. 2nd. ed. Urbana and Chicago: Un. of Illinois, pp. 266–307.

[Sups.83] Suppes, P. 1983, Heuristics and the axiomatic method. In Groner, I., Groner, M. and Bischof, W. F. (eds.), *Methods of Heuristics*. Hillsdale, N. J.: Erlbaum, pp. 79–88.

[Sups.88] Suppes, P. 1988, 'La estructura de las teorias y el analysis de datos', in Suppes, P., *Estudios de Filosofía y Medotologí a de la Ciencia*. Madrid: Alianza, pp. 125–45.

[Sups.02] Suppes, P. 2002, *Representation and Invariance of Scientific Structures*. Stanford: CLI Pu.

[Sups.11] Suppes, P. 2011, Future developments of scientific structures closer to experiment: Response to F. A. Muller. *Synthese* 183 (1): 115–23.

[SylCos.87] Sylvan, R. and da Costa, N.C.A. 1987, Cause as an implication, *Studia Logica* 47 (4), 413–28.

[Sza.64] Szabó, Á. 1964, The transformation of mathematics into deductive science and the beginnings of its foundation on definitions and axioms. *Scripta Mathematica* 27 (1): 27–49.

[Sza.65] Szabó, Á. 1965, Greek dialectic and Euclid's axiomatics. In: Lakatos, I. (ed.), *Problems in the Philosophy of Mathematics* Proceedings of the International Colloquium in the Philosophy of Science, Vol. 1. London: North-Holland: 1–26.

[Sza.78] Szabó, Á. 1978, *The Beginnings of Greek Mathematics*. Dordrecht: D. Reidel (Synthese Historical Library, Vol.17).

[Tar.53] Tarski, A. 1953, *Undecidable Theories*. Amsterdam: North Holland.

[TarGiv.87] Tarski, A. and Givant, S. 1987, *A Formalization of Set Theory without Variables*. Providence: American Mathematical Society.

[Teg.07] Tegmark, M. 2007, Shut up and calculate, arXiv:0709.4024v1

[Tor.81] Toraldo di Francia, G. 1981, *The Investigation of the Physical World*, Cambridge: Cambridge Un. Press.

[Tor.86] Toraldo di Francia, G. 1986, *Le cose e i loro nomi*. Bari: Laterza.

[True.84] Truesdell, C. 1984, *An Idiot's Fugitive Essay on Science: Methods, Criticism, Training, Circumstances*. New York: Springer-Verlag.

[vanF.80] van Fraassen, B. 1980, *The Scientific Image*. Oxford: Clarendon Press.

[vanF.84] van Fraassen, B. 1984. The problem of indistinguishable particles. Reprinted in E. Castellani (ed.) 1998. *Interpreting bodies: classical and quantum objects in modern physics*, Princeton, NY: Princeton University Press, pp. 73–92.

[vanF.89] van Fraassen, B. 1989, *Laws and Symmetry*. Oxford: Clarendon Press.

[vanF.91] van Fraassen, B. 1991, *Quantum Mechanics: An Empiricist View*. Oxford: Clarendon Press.

[vanF.08] van Fraassen, B. 2008, *Scientific Representation*. Oxford: Oxford Un. Press.

[vanH.67] van Heijenoort, J. 1967, *From Frege to Gödel: A Source Book in Mathematical Logic, 1879–1931*. Cambridge: Harvard Un. Press.

[Wei.03] Weingartner, P. (ed.) 2003, *Alternative Logics: Do Sciences Need Them?*. Berlin and Heidelberg: Springer-Verlag.

[WhiRus.10] Whitehead, A. N., and Russell, B. 1910, *Principia Mathematica*, Vol. 1. Cambridge: Merchanf Books.

[Will.70] Williams, M. B. 1970, Deducing the consequences of evolution: a mathematical model. *J. Theoret. Biol.* 29: 343–85.

[Wor.07] Worral, J 2007, "Miracles and Models: Why Reports on the Death of Structural Realism May be Exaggerated", in O'Hare, A. (ed.) *Philosophy of Science (Royal Institute of Pilosophy 61)*, Cambridge: Cambridge University Press pp. 125–54.

[Yan.02] Yandell, B. 2002, *The Honors Class: Hilbert's Problems and Their Solvers*. Natick, MA: A.K.Peters.

[Zah.04] Zahar, E. 2004, Ramseyfication and Structural Realism, *Theoria* 49, pp. 5–30.

[Zer.67] Zermelo, E. 1967/2000, Investigations in the foundations of set theory. In van Heijenoort (ed.), *FromFrege to Gödel: a Source Book in Mathematical Logic, 1879–1931*. Cambridge: Harvard Un. Press.

Index